Groundwater

AWWA MANUAL M21

Second Edition

American Water Works Association

Copyright © 1973, 1989
American Water Works Association
6666 West Quincy Ave.
Denver, CO 80235

Printed in USA

ISBN 0-89867-079-9

Contents

Acknowledgments, v

Introduction vi
    Importance of Groundwater, vi
    Purpose of Manual, vii
    Reference, viii

Chapter 1 The Occurrence and Behavior of Groundwater 1
    Hydrologic Cycle, 1
    Groundwater Occurrence, 2
    Aquifer Characteristics, 4
    Groundwater Movement and Topography, 9
    Land Subsidence, 19
    Reference, 20

Chapter 2 Evaluation of Regional Groundwater Conditions 21
    Suitability of Groundwater Supplies, 21
    Information Sources, 23
    Changes Affecting Evaluation, 24
    Methods for Locating Suitable Groundwater Supplies, 24
    Methods for Monitoring Groundwater Quality, 33
    Field Logistics and Documentation, 35
    References, 38

Chapter 3 Wells—Types, Construction, and Use 39
    Types of Wells and Their Construction, 39
    Common Well Components, 47
    Sanitary Protection, 53
    Reference, 54

Chapter 4 Quantitative Evaluation of Wells 55
    Transmissivity and Storage Coefficient, 55
    Collection of Test Data, 56
    Analysis Procedures, 57
    Well-Field Design, 65
    Well Losses, 68
    Radial-Well Yield, 69
    References, 69

Chapter 5 Use of Computer Models in Groundwater Investigations . . 70
    Models of the Groundwater System, 71
    Reference, 75

Chapter 6 Well Pumps and Pumping **76**
    Pump Classifications, 80
    Operating Conditions, 92
    Pump Selection, 93
    Electric Motor Selection, 96
    Pump Installation, 96

Chapter 7 Common Pump Operating Problems **99**
    Breaking Suction, 99
    Screen Stoppage—Incrustation, 101
    Sand Pumping, 104

Chapter 8 Groundwater Quality and Contamination **106**
    Natural Chemicals in Groundwater, 106
    Groundwater Contamination, 112
    Management of Groundwater Quality, 114
    References, 114

Chapter 9 Groundwater Treatment **115**
    Aeration, 115
    Softening, 116
    Filtration, 117
    Granular Activated Carbon Treatment, 118
    Chlorination, 118
    Fluoridation, 119

Chapter 10 Record Keeping **120**
    Record-Keeping Objectives, 120

Chapter 11 Overview—Design, Construction, and Testing of Wells **122**
    Aquifer Characteristics, 122
    Well Field Evaluation, 125
    Well Field Locations, 129
    Well Design, 131
    Well Development, 135
    Well Testing, 139
    Well Operation and Maintenance, 144

Bibliography, 147

Index, 149

Acknowledgments

The American Water Works Association published the first edition of Manual M21, Ground Water, in 1973. The manual was well received by the water supply industry. In 1984, the AWWA Resources Division recognized that the manual needed to be updated to include new technology. The Resources Division then asked its Groundwater Committee to appoint a task subcommittee to revise the manual. The efforts of that subcommittee resulted in this revision of the groundwater manual.

The members of the task subcommittee who revised this manual include

T.J. Buchanan (Chairman), CH2M Hill, Reston, Va.
R.G. Cousins, Cousins Water Services, Houston, Texas
W.E. Evensen, Salt Lake City Corporation, Salt Lake City, Utah
Laurent McReynolds, Department of Water and Power, Los Angeles, Calif.
Gerald Meyer, US Geological Survey, Reston, Va.
C.E. Nuzman, Groundwater Management, Inc., Kansas City, Kan.
R.T. Sasman, Illinois State Water Survey, Wheaton, Ill.
E.H.A. Stahl, Layne-Northern Company, Lansing, Mich.
C.H. Thompson, URS Corporation, Sacramento, Calif.

Special acknowledgment is given to L. Stephen Lau, University of Hawaii at Manoa, and John F. Mink, Water Resources Consultant, Waipahu, Hawaii, who reviewed the entire manuscript and made substantive recommendations. Many staff members of the US Geological Survey reviewed parts of the manuscript, and many colleagues of the subcommittee members provided guidance and assistance, which is hereby gratefully acknowledged.

Introduction

The quantity of the earth's water resources has remained virtually constant through geologic time. The earth's total water supply is about 1360 million km^3 (Nace 1967), and in general is estimated to be distributed as follows:

Location	Percentage of Total Water
Oceans	97.2
Ice caps and glaciers	2.15
Atmosphere, at sea level	0.001
Subsurface	
Unsaturated zone, including soil moisture	0.005
Groundwater, within a depth of about 800 m	0.31
Groundwater, deep-lying	0.31
Surface	
Freshwater lakes	0.009
Saline lakes and inland seas	0.008
Average instream channels	0.0001
Total	100

IMPORTANCE OF GROUNDWATER

Although the total quantity of the earth's water is unevenly distributed, it is continuously transferring and mixing through the hydrologic cycle. The importance of the earth's groundwater resources can be seen in the comparatively miniscule natural changes occurring within the hydrologic cycle that can severely stress the world's population. Variations in the normally expected quantities of water cause floods and droughts, frequently with devastating environmental and socioeconomic effects, both nationally and internationally.

US Importance

Within this context, groundwater is an extremely important part of the United State's total water supply. Groundwater constitutes about 20 percent of all water used in the United States (excluding hydropower use), and represents the principal source of water for about 48 percent of the total population and about 95 percent of the rural population.

Increases in withdrawal from the nation's groundwater reserves can be expected to occur during the coming years. As demand for groundwater increases, competition for the available supplies will intensify. Several factors will contribute to the anticipated increases, including sharply increased irrigation in the East and West, new water needs for energy production, growing water demands for expanding cities, a declining number of suitable sites for new surface reservoirs, and a desire to establish drought-resistant water supply systems.

The effects of increased withdrawals are expected to be regional in scope. Consequently, protection of this precious resource from contamination requires continued attention, for reasons of public health as well as the costs related to water treatment and aquifer restoration. An ability to understand hydrologic systems and to predict impacts on these systems are essential to the effective development and management of the nation's groundwater resources. Water utility personnel must develop a better understanding of groundwater, and, while no one person can become an expert in all technical fields, a basic understanding and ability to appreciate those fields relating to groundwater is within the grasp of all.

PURPOSE OF MANUAL

The purpose of this manual is to provide general yet up-to-date information that will be of use to all persons concerned with the development, production, and use of groundwater. Personnel working at groundwater utilities will gain an insight into the specialized fields that together make up the study of groundwater. If these individuals understand the problems involved in groundwater development and use and the specialized techniques required to solve these problems and implement the solutions, a great service will be rendered to the utility. The objective of this manual is to provide a basis for such personnel development.

Scope

Chapter 1 discusses the hydrologic cycle and the physical factors that control the behavior and occurrence of groundwater. Knowledge of these factors is fundamental to understanding the processes involved in locating and developing groundwater supplies. In addition to the general physical principles of groundwater, one must learn the significance of particular conditions within the geographic area of interest as they relate to groundwater. These conditions are given in chapter 2, together with methods of quantitative exploration and interpretive examples.

Because of the great variety of local conditions affecting groundwater and individual requirements for this water, a number of water well construction methods have been developed. These methods, the resulting structures, and their use are discussed at some length in chapter 3.

Quantitative area evaluation and test well drilling are preliminary steps that must be taken before adequate testing can be done to determine the amount of groundwater available. The relevant procedures are given in chapter 4. Application of these data to well and well field design is also included.

Mathematical models have become, with the advent of modern computers, valuable tools for managers who make decisions related to the efficient use of groundwater resources. Properly used, mathematical models can assist in the quantitative evaluation of aquifers and the design of groundwater supplies. Chapter 5 contains an overview on the use of computer models in groundwater investigations and in management decisions.

Before a well supply can be put into use, some kind of pumping equipment must be installed. The purpose of a pump is to raise the water from its level in the ground, or other point, to a head where it provides useful pressure. The principles of the general types of pumps used are thoroughly discussed in chapter 6, and items to be considered in selecting the type of equipment and methods for calculating operating costs are also presented.

Each well and pump represent an individual mechanical unit that is subject to operational problems. Common pump operating problems, including breaking of suction, incrustation, and sand pumping, are described in chapter 7. Preventive and routine maintenance are also discussed.

As indicated by the movements of water in the hydrologic cycle, individual droplets of water are subjected to a variety of environments. When water percolates into the ground, it comes in contact with the organic and mineral materials that alter its characteristics continually. These characteristics, known as water quality, are discussed in chapter 8, along with some of the factors contributing to the development of groundwater quality.

Although groundwater is usually quite palatable and acceptable for drinking, its mineral content may make it undesirable for other uses. Altering or removing the minerals that cause problems or adding chemicals that give water desirable attributes is called conditioning—a vast and specialized subject that cannot be thoroughly treated in a manual of this type. There are, however, some basic procedures and processes that are more or less common in systems using groundwater; these are presented in chapter 9.

Chapter 10 includes a discussion of record-keeping programs and data to be collected.

Chapter 11 completes the manual with an overview of the information discussed in previous chapters. Additional information that should be considered when locating, designing, constructing, and testing wells is presented.

Reference

Nace, R.L. Are We Running Out of Water? US Geol. Surv., Circ. No. 536 (1967).

AWWA MANUAL M21

Chapter 1

The Occurrence and Behavior of Groundwater

The groundwater and surface water resources of an area typically are closely related. Such relationships are noted at appropriate places in this and following chapters. This chapter will discuss the occurrence and behavior of groundwater.*

HYDROLOGIC CYCLE

The hydrologic cycle is the constant movement of water above, on, and below the earth's surface (Figure 1-1). The concept of the hydrologic cycle is central to an understanding of the occurrence of water and the development and management of water supplies.

Although the hydrologic cycle has neither a beginning nor an end, it is convenient to discuss its principal features by starting with evaporation from vegetation, exposed moist surfaces, including the land surface, and the oceans. This moisture forms clouds, which return the water to the land surface or oceans in the form of precipitation.

Precipitation and Infiltration

Precipitation occurs in several forms, including rain, snow, and hail, but only rain is considered in this discussion. Initially, rain wets vegetation and other surfaces; it then begins to infiltrate into the ground. Infiltration rates vary widely, depending on land use, the character and moisture content of the soil, and the intensity and duration of precipitation. Rates can vary from as much as 1 in./h (25 mm/h) in mature forests on sandy soils to less than one inch per hour (a few millimetres per hour) in

---

*The material in this chapter has been extracted from a US Geological Survey paper by Heath (1983).

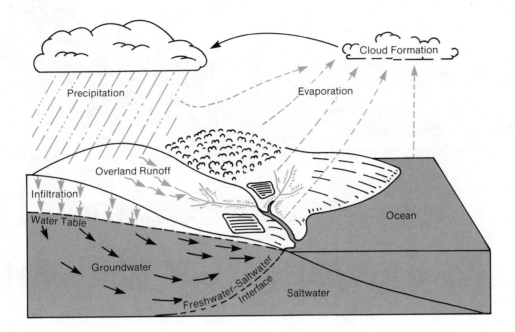

Figure 1-1 The hydrologic cycle.

clayey and silty soils to 0 in./h (0 mm/h) in paved areas. When and if the rate of precipitation exceeds the rate of infiltration, overland flow occurs.

Initial infiltration replaces soil moisture; excess infiltration percolates slowly across the intermediate zone to the zone of saturation (Figure 1-2). From this zone, water moves downward and laterally to sites of groundwater discharge, such as springs on hillsides or seeps in the bottoms of streams and lakes or beneath the ocean. Water reaching streams, both by overland flow and from groundwater discharge, moves to the oceans, where it is again evaporated to continue the cycle.

GROUNDWATER OCCURRENCE

All water beneath the land surface is referred to as underground, or subsurface, water. The equivalent term for water on the land surface is surface water. Underground water occurs in two different zones. One zone, which occurs immediately below the land surface in most areas, contains both water and air and is referred to as the unsaturated zone. The unsaturated zone is almost invariably underlain by a zone in which all interconnected openings are full of water. This zone is referred to as the saturated zone (see Figure 1-2).

Saturated and Unsaturated Zones

Water in the saturated zone is the only underground water that is available to supply wells and springs, and it is the only water to which the name groundwater is correctly applied. Recharge of the saturated zone occurs by percolation of water from the land surface through the unsaturated zone. The unsaturated zone is, therefore, of great importance to groundwater hydrology. This zone may be divided usefully into three parts: the soil zone, the intermediate zone, and the upper part of the capillary fringe.

Soil zone. The soil zone extends from the land surface to a maximum depth of 3 to 5 ft (1 to 2 m). It is the zone that supports plant growth, and it is crisscrossed by

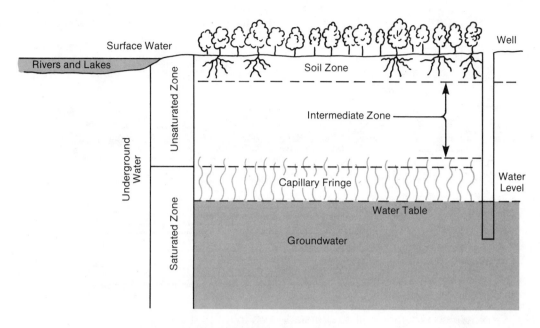

Figure 1-2 Water movement below the earth's surface.

living roots, voids left by decayed roots of earlier vegetation, and animal and worm burrows. The porosity and permeability of the material in this zone tend to be higher than the porosity and permeability of the underlying material.

Intermediate zone. The soil zone is underlain by the intermediate zone, which differs in thickness from place to place, depending on the thickness of the soil zone and the depth to the capillary fringe.

Capillary fringe. The capillary fringe is the subzone between the unsaturated and saturated zones. The capillary fringe results from the attraction between water and rocks. As a result of this attraction, a film of water clings to the surface of rock particles and rises in small-diameter pores against the pull of gravity. Water in the capillary fringe and in the overlying part of the unsaturated zone is under a negative hydraulic pressure—that is, it is under a pressure less than atmospheric (barometric) pressure. The water table, represented by the water level in unused wells, is the level in the saturated zone at which the hydraulic pressure is equal to atmospheric pressure. Below the water table, the hydraulic pressure increases with increasing depth.

Aquifers and Confining Beds

All rocks that underlie the earth's surface can be classified either as aquifers or as confining beds. An aquifer is a rock unit that will yield water in a usable quantity to a well or spring. (In geologic usage, "rock" includes unconsolidated sediments.) A confining bed is a rock unit having very low hydraulic conductivity that restricts the movement of groundwater either into or out of adjacent aquifers (Figure 1-3).

Groundwater occurs in aquifers under two different conditions. Where water only partly fills an aquifer, the upper surface of the saturated zone is free to rise and decline. The water in this type of aquifer is said to be unconfined, and the aquifer is referred to as an unconfined aquifer. Unconfined aquifers are also widely referred to as water-table aquifers. Wells that open to unconfined aquifers are referred to as water-table wells. The water level in these wells indicates the position of the water table in the surrounding aquifer.

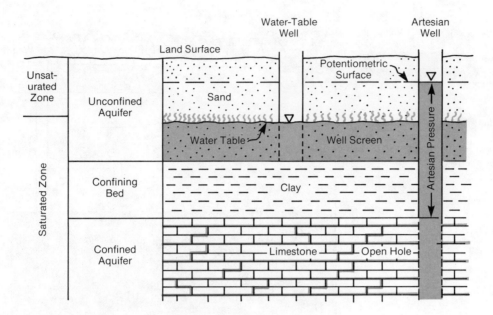

Figure 1-3 Geologic configuration of aquifers and confining beds.

Where water completely fills an aquifer that is overlain by a confining bed, the water in the aquifer is said to be confined. Such aquifers are referred to as confined aquifers or as artesian aquifers. Wells drilled into confined aquifers are referred to as artesian wells. The water level in artesian wells stands at some height above the top of the aquifer but not necessarily above the land surface. If the water level in an artesian well stands above the land surface, the well is a flowing artesian well. The water level in tightly cased wells open to a confined aquifer stands at the level of the potentiometric surface of the aquifer.

AQUIFER CHARACTERISTICS

Numerous parameters are available that are used to describe aquifer characteristics. The most significant of these parameters are porosity, specific yield and specific retention, and heads and gradients.

Porosity

The ratio of openings (voids) to the total volume of a soil or rock is referred to as porosity. Porosity is expressed either as a decimal fraction or as a percentage (Figure 1-4). Thus,

$$n = \frac{V_t - V_s}{V_t} = \frac{V_v}{V_t} \qquad \text{(Eq 1-1)}$$

Where:

n = porosity, as a decimal fraction
V_t = the total volume of a soil or rock sample
V_s = the volume of solids in the sample
V_v = the volume of openings (voids).

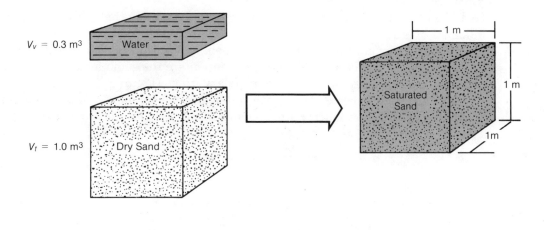

$$\text{Porosity}\ (n) = \frac{\text{Volume of Voids}\ (V_v)}{\text{Total Volume}\ (V_t)} = \frac{0.3\ m^3}{1.0\ m^3} = 0.30$$

Figure 1-4 Definition of porosity.

Table 1-1 Selected Values* of Porosity

Material	Primary Openings	Secondary Openings
Equal-size spheres (marbles)		
Loosest packing	48	—
Tightest packing	26	—
Soil	55	—
Clay	50	—
Sand	25	—
Gravel	20	—
Limestone	10	10
Sandstone (semiconsolidated)	10	1
Granite	—	.1
Basalt (young)	10	1

*Values are given in percent by volume.

If the porosity determined using the above equation is multiplied by 100, the result is porosity expressed as a percentage.

Soils are among the most porous of natural materials, because soil particles tend to form loose clumps and because of the presence of root holes and animal burrows. Porosity of unconsolidated deposits (sands and gravel) depends on the range in grain size (sorting) and on the shape of the rock particles but not on their size. Fine-grained materials tend to be better sorted and, thus, tend to have the highest porosity values. Table 1-1 lists selected values of porosity.

Porosity is important in groundwater hydrology because it reveals the maximum amount of water that a rock can hold when it is saturated. However, it is equally important to know that only a part of this water is available to supply a well or a spring.

Specific Yield and Specific Retention

Hydrologists divide water in ground storage into the portion that will drain under the influence of gravity, which is called specific yield, and the portion that is retained as a film on rock surfaces and in very small openings, which is called specific retention. The physical forces that control specific retention are the same forces controlling the thickness and moisture content of the capillary fringe (Figure 1-5).

Specific yield indicates how much water is available for man's use, and specific retention indicates how much water remains in the rock after it is drained by gravity. Thus,

$$n = S_y + S_r \qquad \text{(Eq 1-2)}$$

$$S_y = \frac{V_d}{V_t} \text{ and } S_r = \frac{V_r}{V_t} \qquad \text{(Eq 1-3)}$$

Where:

n = porosity
S_y = specific yield
S_r = specific retention
V_d = the volume of water that drains from a total volume of V_t
V_r = the volume of water retained in a total volume of V_t
V_t = total volume of a soil or rock sample.

Table 1-2 lists selected values of porosity, specific yield, and specific retention.

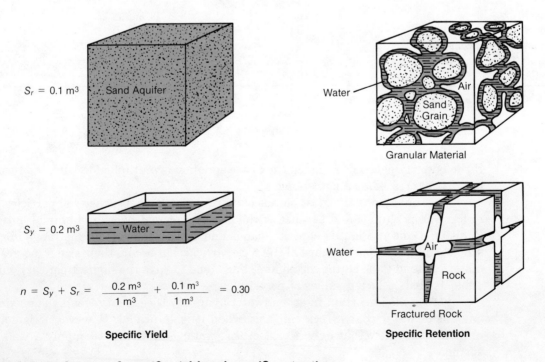

Figure 1-5 Definition of specific yield and specific retention.

Table 1-2 Selected Values* of Porosity, Specific Yield, and Specific Retention

Material	Porosity	Specific Yield	Specific Retention
Soil	55	40	15
Clay	50	2	48
Sand	25	22	3
Gravel	20	19	1
Limestone	20	18	2
Sandstone (semiconsolidated)	11	6	5
Granite	.1	.09	.01
Basalt (young)	11	8	3

*Values are given in percent by volume.

Heads and Gradients

The depth to the water table has an important effect on the way in which the land surface is used and on the development of water supplies from unconfined aquifers. Where the water table is at a shallow depth, the land may become waterlogged during wet weather and unsuitable for residential and other uses. Where the water table is at great depth, the cost of constructing wells and pumping water for domestic needs may be prohibitively expensive.

The direction that the water table slopes is important because it indicates the direction of groundwater movement (Figure 1-6). The position and the slope of the water table (or of the potentiometric surface of a confined aquifer) is determined by measuring the position of the water level in wells from a fixed point (a measuring point). To use these measurements to determine the slope of the water table, the position of the water table at each well must be determined relative to a datum plane that is common to all the wells. The datum plane most widely used is the National Geodetic Vertical Datum of 1929, also commonly referred to as sea level.

Total head. If the depth to water in a nonflowing well is subtracted from the altitude of the measuring point, the result is the total head at the well. Total head, as defined in fluid mechanics, encompasses elevation head, pressure head, and velocity

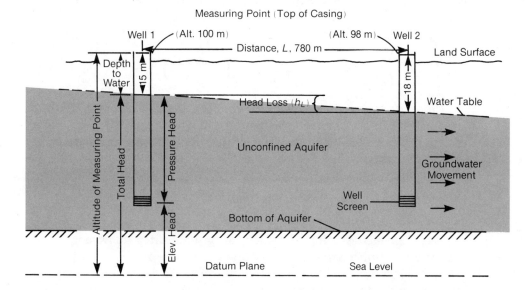

Figure 1-6 Definition of heads and gradients.

head. Because groundwater moves relatively slowly, velocity head can be ignored. Therefore, the total head at an observation well involves only two components: elevation head and pressure head. Note that groundwater moves in the direction of decreasing total head, which may or may not be in the direction of decreasing pressure head.

The equation for total head h_t is

$$h_t = z + h_p \qquad \text{(Eq 1-4)}$$

Where:

z = elevation head, and is the distance from the datum plane to the point where the pressure head h_p is determined.

Hydraulic gradient. All other factors being constant, the rate of groundwater movement depends on the hydraulic gradient. The hydraulic gradient is the change in head per unit of distance in a given direction. If the direction is not specified, it is understood to be in the direction in which the maximum rate of decrease in head occurs.

As an example, if the movement of groundwater is in the plane shown in Figure 1-6, that is, if it moves from well 1 to well 2, the hydraulic gradient can be calculated from the information given on the drawing. The hydraulic gradient is h_L/L, where h_L is the head loss between wells 1 and 2, and L is the horizontal distance between them. Using the measurements given in Figure 1-6, this can be expressed as

$$\frac{h_L}{L} = \frac{(100 \text{ m} - 15 \text{ m}) - (98 \text{ m} - 18 \text{ m})}{780 \text{ m}} = \frac{85 \text{ m} - 80 \text{ m}}{780 \text{ m}} = \frac{5 \text{ m}}{780 \text{ m}} \qquad \text{(Eq 1-5)}$$

When the hydraulic gradient is expressed in consistent units, as it is in the above example in which both the numerator and the denominator are in metres, any other consistent units of length can be substituted without changing the value of the gradient. Thus, a gradient of 5 ft/780 ft is the same as a gradient of 5 m/780 m. It is also relatively common to express hydraulic gradients in inconsistent units, such as metres per kilometre or feet per mile. A gradient of 5 m/780 m can be converted to metres per kilometre as follows:

$$\left(\frac{5 \text{ m}}{780 \text{ m}}\right) \times \left(\frac{1000 \text{ m}}{\text{km}}\right) = 6.4 \text{ m/km}$$

Calculating groundwater movement and hydraulic gradient. Both the direction of groundwater movement and the hydraulic gradient can be determined if the following data are available for three wells located in any triangular arrangement, such as that shown on Figure 1-7:
- the relative geographic position of the wells,
- the distance between the wells, and
- the total head at each well.

Steps in determining direction of groundwater movement and hydraulic gradient are outlined below and illustrated in Figure 1-8.

1. Identify the well that has the intermediate water level, that is, neither the highest head nor the lowest head.

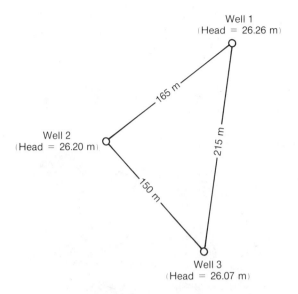

Figure 1-7 Example well location to be used in determining direction of groundwater movement and hydraulic gradient.

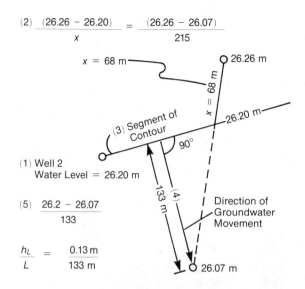

Figure 1-8 Steps in determining direction of groundwater movement and hydraulic gradient.

2. Calculate the position between the well having the highest head and the well having the lowest head at which the head is the same as that in the intermediate well.

3. Draw a straight line between the intermediate well and the point identified in step 2 as being between the well having the highest head and that having the lowest head. This line represents a segment of the water-level contour along which the total head is the same as that in the intermediate well.

4. Draw a line perpendicular to the water-level contour and through either the well with the highest head or the well with the lowest head. This line parallels the direction of groundwater movement.

5. Divide the difference between the head of the well and that of the contour by the distance between the well and the contour. The answer is the hydraulic gradient.

GROUNDWATER MOVEMENT AND TOPOGRAPHY

It is desirable, wherever possible, to determine the position of the water table and the direction of groundwater movement. Methods to determine direction of groundwater movement were discussed above. However, in most areas, general but very valuable conclusions about the direction of groundwater movement can be derived from observations of land topography.

Gravity is the dominant driving force in groundwater movement. Under natural conditions, groundwater moves downhill until, in the course of its movement, it reaches the land surface at a spring or through a seep along the side or bottom of a stream channel or estuary. Thus, groundwater in the shallowest part of the saturated zone moves from interstream areas toward streams or the coast.

If minor land-surface irregularities are ignored, the slope of the land surface is also toward streams or the coast. In effect, the water table usually is a subdued replica of the land surface (Figure 1-9).

The potentiometric surface of confined aquifers, like the water table, also slopes from recharge areas to discharge areas. Shallow confined aquifers, which are

10 GROUNDWATER

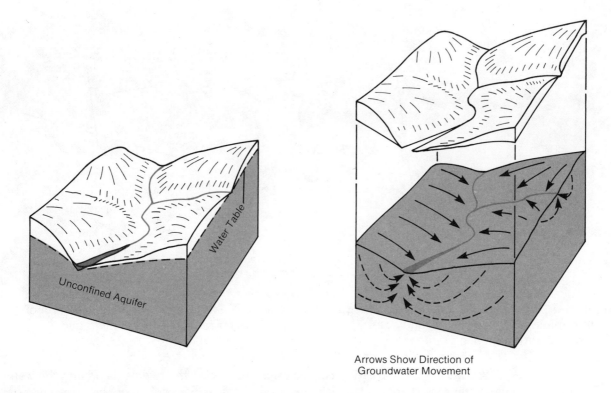

Figure 1-9 Groundwater movement as it relates to topography.

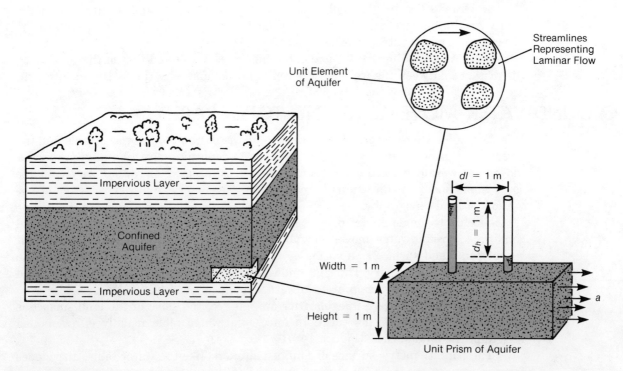

Figure 1-10 Definition of hydraulic conductivity.

relatively common along the Atlantic Coastal Plain, share both recharge and discharge areas with the surficial unconfined aquifers. However, this sharing may not exist in deeper confined aquifers, where the principal recharge areas are generally in outcrop areas near the western border of the Coastal Plain. The discharge areas for deeper confined aquifers are generally near the heads of estuaries that occur along major streams. Thus, movement of water through these aquifers is in a general west to east direction, where it has not been modified by withdrawals.

In the western part of the conterminous United States and especially in the alluvial basins region,* conditions are more variable than those described above. In this area, streams flowing from mountain ranges onto alluvial plains lose water to the alluvial deposits. Thus, groundwater in the upper part of the saturated zone flows down the valleys and at an angle away from the streams.

Hydraulic Conductivity

The factors controlling groundwater movement were first expressed by Henry Darcy, a French engineer, in 1856. Darcy's law is as follows:

$$Q = KA \left(\frac{dh}{dl}\right) \tag{Eq 1-6}$$

Where:

Q = the quantity of water per unit of time

K = the hydraulic conductivity, which depends on the size and arrangement of the water-transmitting openings (pores and fractures) and on the dynamic characteristics of the fluid (water), such as kinematic viscosity, density, and the strength of the gravitational field

A = the cross-sectional area, at a right angle to the flow direction, through which the flow occurs

dh/dl = the hydraulic gradient†

Because the quantity of water Q is directly proportional to the hydraulic gradient dh/dl, groundwater flow is said to be laminar, that is, water particles tend to follow discrete streamlines and not to mix with particles in adjacent streamlines (Figure 1-10).

If Eq 1-6 is rearranged to solve for K, the following is obtained:

$$K = \frac{Qdl}{Adh} = \frac{(m^3/d)(m)}{(m^2)(m)} = \frac{m}{d} \tag{Eq 1-7}$$

---

*The alluvial basins region occupies a discontinuous area that extends from the Puget Sound–Williamette Valley area of Washington and Oregon to west Texas. This region consists of alternating basins or valleys and mountain ranges.

†Where hydraulic gradient is discussed as an independent entity, as it is in the subsection "Heads and Gradients," it is shown symbolically as h_L/L and is referred to as head loss per unit of distance. Where hydraulic gradient appears as one of the factors in an equation, as it does in Eq 1-6, it is shown symbolically as dh/dl to be consistent with other groundwater literature. The gradient dh/dl indicates that the unit distance is reduced to as small a value as one can imagine, in accordance with the concepts of differential calculus.

Thus, the units of hydraulic conductivity are those of velocity (or distance divided by time). It is important to note from Eq 1-7, however, that the factors involved in the definition of hydraulic conductivity include the volume of water Q that will move in a unit of time (commonly, a day) under a unit hydraulic gradient (such as a metre per metre) through a unit area (such as a square metre). These factors are illustrated in Figure 1-10. Expressing hydraulic conductivity in terms of a unit gradient rather than an actual gradient at some place in an aquifer permits ready comparison of values of hydraulic conductivity for different rocks.

Hydraulic conductivity in rock. The hydraulic conductivity of rocks ranges through 12 orders of magnitude (Figure 1-11). There are few physical parameters whose values range so widely. Hydraulic conductivity is not only different in different types of rocks but may also be different from place to place in the same rock. If the hydraulic conductivity is essentially the same in any area, the aquifer in that area is said to be homogeneous. If, on the other hand, the hydraulic conductivity differs from one part of the area to another, the aquifer is said to be heterogeneous.

Hydraulic conductivity may also be different in different directions at any place in an aquifer. If the hydraulic conductivity is essentially the same in all directions, the aquifer is said to be isotropic. If it is different in different directions, the aquifer is said to be anisotropic.

Although it is convenient in many mathematical analyses of groundwater flow to assume that aquifers are both homogeneous and isotropic, such aquifers are rare, if

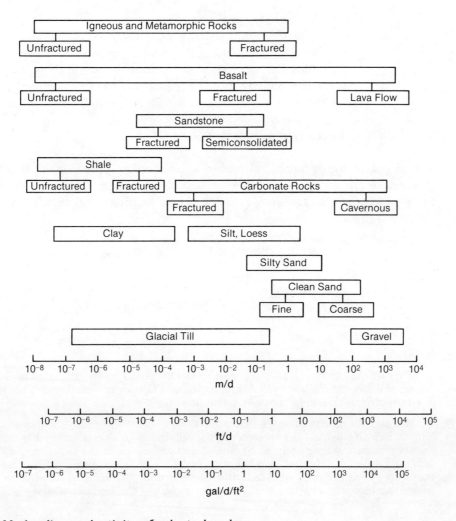

Figure 1-11 Hydraulic conductivity of selected rocks.

they exist at all. The condition most commonly encountered is for hydraulic conductivity in most rocks and especially in unconsolidated deposits and in flat-lying consolidated sedimentary rocks to be larger in the horizontal direction than it is in the vertical direction.

It should be noted that hydraulic conductivity replaces the term "field coefficient of permeability" and should be used in referring to the water-transmitting characteristic of material in quantitative terms. It is still common practice to refer in qualitative terms to "permeable" and "impermeable" material.

Capillarity and Unsaturated Flow

Most recharge of groundwater systems occurs during the percolation of water across the unsaturated zone (see Figure 1-2). This movement of water is controlled by both gravitational and capillary forces.

Capillarity results from two forces: the mutual attraction (cohesion) between water molecules and the molecular attraction (adhesion) between water and different solid materials. Most pores in granular materials are of capillary size. As a result, water is pulled upward into a capillary fringe above the water table to a height h_c above the water level in the same manner that water is pulled up into a column of sand whose lower end is immersed in water (Figure 1-12, Table 1-3).

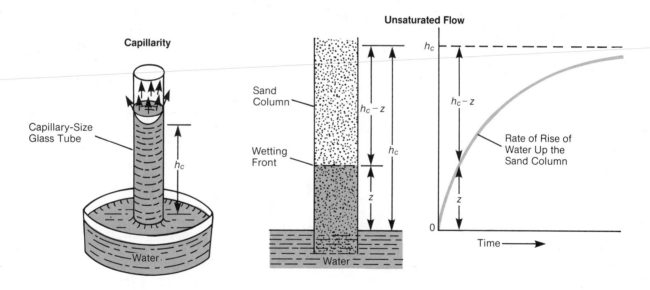

Figure 1-12 Definition of capillarity and unsaturated flow.

Table 1-3 Approximate Height of Capillary Rise h_c in Granular Materials

Material	Rise *in.*
Sand	
Coarse	5
Medium	10
Fine	15
Silt	40

A steady-state* flow of water in the unsaturated zone can be determined from a modified form of Darcy's law. Steady-state unsaturated flow Q is proportional to the effective hydraulic conductivity K_e, the cross-sectional area A through which the flow occurs, and gradients due to both capillary forces and gravitational forces. Thus,

$$Q = K_e A \left(\frac{h_c - z}{z}\right) \pm \left(\frac{dh}{dl}\right) \qquad \text{(Eq 1-8)}$$

Where:

Q = the quantity of water
K_e = the hydraulic conductivity under the degree of saturation existing in the unsaturated zone
A = the cross-sectional area through which flow occurs
$(h_c - z)/z$ = the gradient due to capillary (surface tension) forces
dh/dl = the gradient due to gravity.

The plus/minus sign is related to the direction of movement—plus for downward and minus for upward. For movement in a vertical direction, either up or down, the gradient due to gravity is 1/1, or 1. For lateral (horizontal) movement in the unsaturated zone, the term for the gravitational gradient can be eliminated.

The capillary gradient at any time depends on the length of the water column z supported by capillarity in relation to the maximum possible height of capillary rise h_c (see Figure 1-12). For example, if the lower end of a sand column is suddenly submerged in water, the capillary gradient is at a maximum, and the rate of rise of water is fastest. As the wetting front advances up the column, the capillary gradient declines, and the rate of rise decreases.

The capillary gradient can be determined from tensiometer measurements of hydraulic pressures. To determine the gradient, it is necessary to measure the negative pressure h_p at two levels in the unsaturated zone, as Figure 1-13 shows. The equation for total head h_t is

$$h_t = z + h_p \qquad \text{(Eq 1-9)}$$

Where:

z = the elevation of a tensiometer.

Substituting values in this equation for tensiometer number 1, the following is obtained:

$$h_t = 32 + (-1) = 32 - 1 = 31 \text{ m} \qquad \text{(Eq 1-10)}$$

The total head at tensiometer number 2 is 26 m. The vertical distance between the tensiometers is 32 m minus 28 m, or 4 m. Because the combined gravitational and capillary hydraulic gradient equals the head loss divided by the distance between tensiometers, the gradient is

---

*Steady state in this context refers to a condition in which the moisture content remains constant, as it would, for example, beneath a waste-disposal pond whose bottom is separated from the water table by an unsaturated zone.

$$\frac{h_L}{L} = \frac{h_{t(1)} - h_{t(2)}}{z_{(1)} - z_{(2)}} = \frac{31 - 26}{32 - 28} = \frac{5 \text{ m}}{4 \text{ m}} = 1.25 \qquad \text{(Eq 1-11)}$$

This gradient includes both the gravitational gradient dh/dl and the capillary gradient $(h_c - z)/z$. Because the head in tensiometer number 1 exceeds that in tensiometer number 2, it is known that flow is vertically downward and that the gravitational gradient is 1/1, or 1. Therefore, the capillary gradient is 0.25 m/m (1.25 − 1.00).

The effective hydraulic conductivity K_e is the hydraulic conductivity of material that is not completely saturated. It is thus less than the (saturated) hydraulic conductivity K_s for the material. Figure 1-14 shows the relation between degree of saturation and the ratio of saturated and unsaturated hydraulic conductivity for coarse sand. The hydraulic conductivity K_s of coarse sand is about 60 m/d.

Transmissivity

The capacity of an aquifer to transmit water of the prevailing kinematic viscosity is referred to as its transmissivity. The transmissivity T of an aquifer is equal to the hydraulic conductivity of the aquifer multiplied by the saturated thickness of the aquifer. Thus,

$$T = Kb \qquad \text{(Eq 1-12)}$$

Where:

T = transmissivity
K = hydraulic conductivity
b = aquifer thickness.

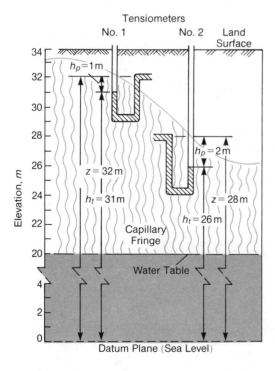

Figure 1-13 Determining capillary gradient from tensiometer measurements of hydraulic pressures.

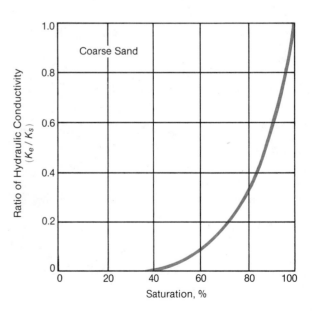

Figure 1-14 Relation between degree of saturation and the ratio of saturated and unsaturated hydraulic conductivity for coarse sand.

As is the case with hydraulic conductivity, transmissivity is also defined in terms of a unit hydraulic gradient.

If Eq 1-12 is combined with Darcy's law (see the subsection "Hydraulic Conductivity"), the result is an equation that can be used to calculate the quantity of water q moving through a unit width w of an aquifer. Darcy's law is

$$q = KA\left(\frac{dh}{dl}\right) \tag{Eq 1-13}$$

Expressing area A as bw,

$$q = Kbw\left(\frac{dh}{dl}\right) \tag{Eq 1-14}$$

Next, substituting transmissivity T for Kb,

$$q = Tw\left(\frac{dh}{dl}\right) \tag{Eq 1-15}$$

Eq 1-15 modified to determine the quantity of water Q moving through a large width W of an aquifer is

$$Q = TwW\left(\frac{dh}{dl}\right) \tag{Eq 1-16}$$

or, if it is recognized that T applies to a unit width w of an aquifer, this equation can be stated more simply as

$$Q = TW\left(\frac{dh}{dl}\right) \tag{Eq 1-17}$$

If Eq 1-17 is applied to Figure 1-15, the quantity of water flowing from the right-hand side of the drawing can be calculated by using the values shown as follows:

$$T = Kb = \frac{50\ \text{m}}{\text{d}} \times \frac{100\ \text{m}}{1} = 5000\ \text{m}^2/\text{d} \tag{Eq 1-18}$$

$$Q = TW\left(\frac{dh}{dl}\right) = \frac{5000\ \text{m}^2}{\text{d}} \times \frac{1000\ \text{m}}{1} \times \frac{1\ \text{m}}{1000\ \text{m}} = 5000\ \text{m}^3/\text{d} \tag{Eq 1-19}$$

Equation 1-17 is also used to calculate transmissivity, where the quantity of water Q discharging from a known width of aquifer can be determined as, for example, with streamflow measurements. Rearranging terms, the following is obtained:

$$T = \frac{Q}{W}\left(\frac{dl}{dh}\right) \tag{Eq 1-20}$$

The units of transmissivity, as the preceding equation demonstrates, are

$$T = \frac{(\text{m}^3/\text{d})(\text{m})}{(\text{m})(\text{m})} = \frac{\text{m}^2}{\text{d}} \tag{Eq 1-21}$$

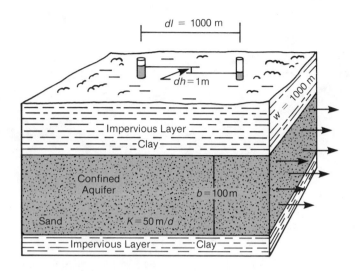

Figure 1-15 Definition of transmissivity.

Calculating transmissivity. Figure 1-16 illustrates the hydrologic situation that permits calculation of transmissivity through the use of stream discharge. The calculation can be made only during dry-weather (baseflow) periods, when all water in the stream is derived from groundwater discharge. For the purpose of this example, the following values are assumed:

Average daily flow at stream-gauging station A: 2.485 m³/s
Average daily flow at stream-gauging station B: 2.355 m³/d
Increase in flow due to groundwater discharge: 0.130 m³/s
Total daily groundwater discharge to stream: 11,232 m³/d
Discharge from half of aquifer (one side of the stream): 5616 m³/d
Distance x between stations A and B: 5000 m
Average thickness of aquifer b: 50 m
Average slope of the water table dh/dl determined from measurements in the observation wells: 1 m/2000 m

Using Eq 1-20

$$T = \frac{Q}{W} \times \frac{dl}{dh} = \frac{5616 \text{ m}^3}{\text{d} \times 5000 \text{ m}} \times \frac{2000 \text{ m}}{1 \text{m}} = 2246 \text{ m}^2/\text{d} \qquad (\text{Eq 1-22})$$

The hydraulic conductivity is determined from Eq 1-12 as follows:

$$K = \frac{T}{b} = \frac{2246 \text{ m}^2}{\text{d} \times 50 \text{ m}} = 45 \text{ m/d} \qquad (\text{Eq 1-23})$$

Because transmissivity depends on both K and b, its value differs in different aquifers and from place to place in the same aquifer. Estimated values of transmissivity for the principal aquifers in different parts of the United States range from less than 1 m²/d for some fractured sedimentary and igneous rocks to 100,000 m²/d for cavernous limestones and lava flows.

Finally, transmissivity replaces the term "coefficient of transmissibility" because, by convention, an aquifer is transmissive and the water in it is transmissible.

18 GROUNDWATER

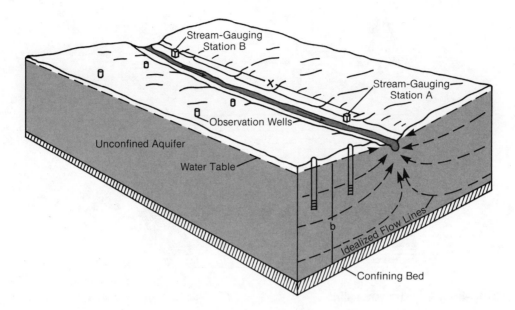

Figure 1-16 Calculation of transmissivity using stream discharge.

Storage Coefficient

The abilities (capacities) of water-bearing materials to store and transmit water are their most important hydraulic properties. These properties are given either in terms of a unit cube of the material or in terms of a unit prism of an aquifer, depending on the intended use. These abilities, as they relate to the two units of measurement, are presented below.

Property	Unit Cube of Material	Unit Prism of Aquifer
Transmissive capacity	Hydraulic conductivity K	Transmissivity T
Available storage	Specific yield S_y	Storage coefficient S

The storage coefficient S is defined as the volume of water that an aquifer releases from or takes into storage per unit surface area of the aquifer per unit change in head. The storage coefficient is a dimensionless unit, as the following equation shows, in which the units in the numerator and the denominator cancel.

$$S = \frac{\text{volume of water}}{(\text{unit area})(\text{unit head change})} = \frac{m^3}{(m^2)(m)} = \frac{m^3}{m^3} \quad \text{(Eq 1-24)}$$

The size of the storage coefficient depends on whether the aquifer is confined or unconfined (Figure 1-17). If the aquifer is confined, the water released from storage when the head declines comes from expansion of the water and from compression of the aquifer. Relative to a confined aquifer, the expansion of a given volume of water in response to a decline in pressure is very small. In a confined aquifer having a porosity of 0.2 and containing water at a temperature of about 59°F (15°C), expansion of the water alone releases about 3×10^{-7} m^3 of water per cubic metre of aquifer per metre of decline in head.

To determine the storage coefficient of an aquifer due to expansion of the water, it is necessary to multiply the aquifer thickness by 3×10^{-7}. Thus, if only the expan-

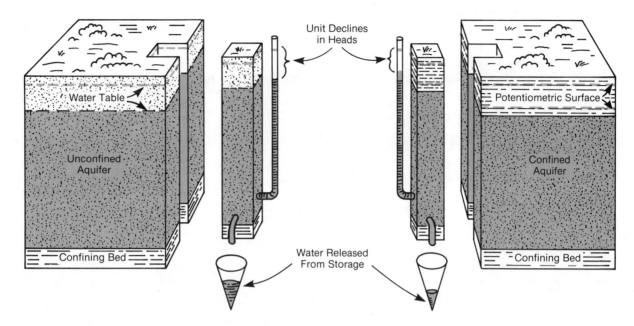

Figure 1-17 Definition of storage coefficient.

sion of water is considered, the storage coefficient of an aquifer 33-ft (100-m) thick would be 3×10^{-5}. The storage coefficient of most confined aquifers ranges from about 10^{-5} to 10^{-3}. The difference between these values and the value due to expansion of the water is attributed to compression of the aquifer.

LAND SUBSIDENCE

A significant consequence of groundwater development can be downward movement of the land surface, which is called subsidence. Planning development of groundwater in some areas needs to include consideration of the impact of land-surface subsidence.

Estimating Subsidence

Reasonable estimates of the magnitude of subsidence in areas subject to flooding either by tidal inundation or alteration of surface drainage need to be made prior to development. If movement along faults (activation or acceleration) leads to structural damage and if the movement is related to man-caused subsidence, estimates of the magnitude of subsidence also need to be made.

The information necessary for estimating probable subsidence includes amount of compressible material, stress on the system (water-level change), and degree of compressibility of the subsurface material. Unfortunately, the information needed is not available in sufficient detail in most areas.

The most readily available of the necessary data is the amount of compressible material in the subsurface. Such data may be obtained from evaluation of logs of well test holes. Data on water-level changes can be used to make estimates of pressure change (stress change) at various depths for various time intervals.

Data on the degree of compressibility of the subsurface material are less readily available than either of the other factors used to predict subsidence. Laboratory values of compressibility determined from tests of cores have been used with limited success. The expense of obtaining undisturbed cores and the difficulty in obtaining representative cores preclude their use for regional appraisal. Where subsidence has been well documented, subsidence data may be coupled with historic stress changes and the amount of compressible material to determine compressibility.

Reference

HEATH, R.C. Basic Ground-water Hydrology. US Geol. Surv. Water-Supply Paper 2220, US Govt. Printing Ofce., p. 4–28 (1983).

AWWA MANUAL M21

Chapter **2**

Evaluation of Regional Groundwater Conditions

The evaluation of regional groundwater conditions and the potential for resource development should be based on the following factors:
- the quantity and quality of water required;
- the scarcity of water resources versus regional demand (in other words, the cost of water in a region);
- the nature and density of existing or likely future pollution sources and the effectiveness of regulatory controls on these sources;
- the level of previous groundwater development and hydrogeologic or groundwater pollution investigation; and
- the understanding of long-term land uses affecting quality and quantity of groundwater recharge.

These factors along with special considerations, such as conveyance distances or treatment pertinent to a given development project, may be combined when determining the costs involved for resource development evaluation.

SUITABILITY OF GROUNDWATER SUPPLIES

Past perceptions of assumed groundwater purity have created expectations of consistently reliable groundwater sources. Unfortunately, groundwater sources can no longer be assumed as safe. Water quality concerns, such as severe mineralization and contamination from a variety of sources, have made groundwater supplies unsuitable for human use in some instances. Questionable land use practices are also causes for concern.

Water Quality Considerations

Traditionally, materials that have affected the value of a groundwater supply have included naturally occurring minerals, such as dissolved inorganic salts. Within many regions, the mineral quality of groundwater is fairly uniform at certain depth ranges. Some investigators would argue that a significant relation exists between mineral quality and depth. It may be thought that drilling deeper is the simple solution to the problem of severe mineralization. However, the mineral quality of groundwater commonly declines with depth in deeper zones. Groundwater quality in many sedimentary basins, where the older (deeper) sediments were deposited in a marine environment, can change very abruptly in mineral content as a well penetrate into the marine sediments. Poor-quality water can be drawn upward after production begins, even if a production well does not penetrate a saline zone. Similarly, operation of coastal production wells can induce saltwater intrusion into freshwater aquifers.

Today, other forms of contamination exist. The indications are that synthetic and naturally occurring organic compounds, plus refined minerals and heavy metals, must be considered when evaluating the development potential of a groundwater resource.

Published information on the hydrogeology of an area will usually provide a description of the general water quality conditions. However, water quality can change with time (especially in the shallower groundwater zones) and local variations will not appear in the published literature. Therefore, some exploratory work will have to be done to provide needed details regarding water quality.

Pollution control. The types of waste generated within local areas, methods of handling and disposing of the waste, the likelihood of accidental and/or unreported spills and leaks, and the hydrogeology of intervening materials are all important considerations when evaluating groundwater quality. It should be determined whether or not proposed well development will be drawing water that has been, is now, or will be impacted by groundwater contamination.

Today, urban-area groundwater-resource developers exercise caution when evaluating groundwater sites, because testing for contaminants has become more the norm than the exception. Local and state statutes and regulations are beginning to impose stringent testing requirements on developers and water suppliers. Such testing requirements make knowledge of organic and inorganic water quality parameters available, which in turn may allow treatment costs to be considered during initial groundwater surveys. This information can be used to support conclusions or recommendations that an otherwise unsuitable water resource has development potential.

Land Use

Regional conditions, especially land use, contributing to groundwater quality must be considered when evaluating groundwater sources. After an adequate picture of the hydrogeology and groundwater flow patterns has been developed for a proposed site, past, present, and current land uses in critical areas around the site should be evaluated to determine the potential for groundwater contamination.

Even apparent benign land uses, such as agriculture, may have very significant impacts on groundwater quality. Residential land use may also impact groundwater quality. For example, septic tanks receive a variety of household chemicals; the field lines from these tanks carry leachate that can cause these chemicals to migrate, with little change, into an aquifer. Pollution source areas, which are up-gradient of the proposed site, are of most concern.

As land uses change, an aquifer with near pristine quality at the time of development may deteriorate. Consequently, the resource developer must obtain and maintain a good understanding of urban and industrial growth and zoning of the area associated with the groundwater supply. Today, simulation models, which depict the long-term effects of a proposed water-well development and provide documentation of any land-use changes that may affect the local hydrology (for example, industrial development displacing irrigated agriculture or urbanization paving that reduces recharge), will define up-gradient sources. This modeling can take the form of analytical solutions for simple cases or numerical computer codes for more complex cases where a high degree of accuracy is needed.

Groundwater Modeling

Judgment is required for all methods of groundwater modeling, since there is usually a lack of certainty about flow patterns and possible changes in flow patterns. In addition, flow of discharged material in the unsaturated zone (discussed in chapter 1) is not governed by groundwater gradients, and contaminants can move contrary to prevailing groundwater flow. Groundwater development should be down gradient an appropriate distance from potential threats to the water quality. It may be appropriate to analyze for more compounds at greater frequency, or to establish early-warning monitoring wells at various depths, or to use other tools that will create certainty as to the future groundwater quantity and quality.

INFORMATION SOURCES

The careful use of published literature can reduce some expensive field study and should form the basis of any hydrogeologic investigation.

Published Reports

Sources of information to begin a survey of regional groundwater conditions include utility-commission reports from existing public suppliers, reports from local drillers, and published reports from government agencies. Some states consider drillers' logs and well-completion details as confidential. However, requirements of public good and need to know might be used to negotiate access to these records for specific projects.

Published reports on groundwater resources are available covering various geographic regions, including major basin areas and state, county, and local regions. For studies within the United States, a good place to begin the literature search is the Summary Appraisals of the Nation's Ground-Water Resources (1978–1982). An individual report (one chapter) covering the region of interest can be obtained separately or a compilation of all 20 reports is offered by Todd (1984). These reports provide an excellent summary of reliable information on quantity and quality of available groundwater. By examining the supporting references, which are cited for specific localities, data can be arranged into a composite array that serves to complete the available facts.

Valuable information regarding contamination sources may be obtained from state and federal programs that administer hazardous waste or effluent legislative programs. These programs characterize waste and monitor groundwater quality, as preliminary steps before granting operating permits for groundwater wells.

Caution should be used in any investigation, however, because even fairly site-specific reports are necessarily general in nature, and many local details may be omitted. Local conditions may differ from the regional average, and if good prospects for resource development are rare, one may have to explore areas that do not at first appear attractive in the general reports.

In most parts of the United States, knowledge regarding quantity of groundwater resources is reasonably complete. In less-developed regions, available information will usually be limited. For studies in these areas, a reconnaissance-type survey would suffice, especially as an initial effort. In highly developed regions, a wealth of information will already exist, and a very detailed evaluation will be required. Where the available groundwater resources have often been developed, the groundwater survey should be tailored to only fill gaps in knowledge about the specific locations to be developed within the region.

Application

The water developer should establish that his/her data and analyses adequately demonstrate the feasibility of new groundwater development. The development should be shown to not infringe on existing water rights or competing water uses. This will reduce the chance of disputes arising, and help provide for resolution of any disputes that do arise. The water developer should be able to demonstrate "sustainable yield" and also show that the side effects of pumping, such as land subsidence or saltwater intrusion, or interference with competing water or competing land users would be negligible. Although investigation of the hydrogeology will be central to the evaluation, a more up-to-date and meaningful study will include current and forecasted water quality.

CHANGES AFFECTING EVALUATION

Changes in land uses and changes in the public's needs for water of a given quality and quantity make it essential to have reliable measurements to evaluate any regional groundwater system.

Actual and Perceived Changes

Changes pertaining to any groundwater resource can be actual or perceived. Actual deterioration of water quantity and quality can have grave consequences for those dependent on the water resource. Changes other than those measured in the field can often be of great concern. Perceived changes may be due to the availability of increased levels of analytical detection for determining levels of compounds not previously measurable; governmental priority shifts toward protection of natural resources, including aquifer classifications; new toxicological data or reinterpretation of existing toxicology; and over-simplification and integration of facts regarding groundwater quality into conservative measures toward the safe use of all water resources.

Consequently, periodic, routine examinations, as well as nonroutine examinations of the groundwater system are necessary. The need to conduct at least periodic reevaluations appears reasonable, and the developer of the resource should anticipate these reevaluations in order to minimize unexpected responses from the water-resource-development agency.

METHODS FOR LOCATING SUITABLE GROUNDWATER SUPPLIES

The goal of groundwater exploration is to locate productive aquifers that yield sustainable high-quality water. The production level and quality required will guide exploration efforts, but interest will generally center around locating large deposits of the following:

- unconsolidated sands and gravels of alluvial or glacial origin;
- sandstones and conglomerates;
- limestones and dolomites; and
- porous or fractured volcanic rocks.

It should be noted that many other types of rock can yield small quantities of water to meet the domestic needs of a single household.

Depending on the region of interest, the starting place of the exploration effort will vary. In undeveloped regions, reconnaissance-type interpretation of aerial photos, LANDSAT images, and maps may be the first step. Drainage density measurements as an indicator of recharge efficiency can also be useful. In regions with uniform rainfall, areas having the greatest surface-drainage capacity will have the least amount of groundwater recharge and would be of least interest for groundwater development. Cross-sectional maps may have to be drawn from surface geologic information and existing well logs. These maps will aid in determining the location, depth, and thickness of favorable aquifers.

In many regions, general cross sections and hydrogeologic interpretations will already exist, along with well logs and production information from existing wells. In these cases, favorable aquifers will have already been identified, and exploration efforts can proceed in order to fill in details regarding the local hydrogeologic environment. This can be accomplished using surface geophysical methods, borehole geophysical methods, aquifer testing, exploratory drilling, and lithologic logging.

Surface Geophysical Methods

Surface geophysical methods, principally electrical resistivity and seismic reflection and refraction, can be used to provide a more complete picture of subsurface structure, given some prior knowledge obtained from surface geology and borehole logs. A general model of the subsurface geology should be available to provide the basis for a proper interpretation of the surface geophysical data. The information generated can be used to locate sites for further investigation with test drilling. In rare circumstances of very well-defined subsurface geology and well-documented groundwater movement, some contaminant plumes may be projected using geophysical methods of measurement.

Successful application of any surface geophysical method depends on the presence of sharp discontinuities in the physical properties affecting the measurements. The detectable physical properties only provide indirect estimates of the hydrogeologic properties of interest, with the accuracy of such estimates being dependent on how closely these physical properties relate to each other. In addition, the structural configurations amenable to investigation must be relatively simple. A comprehensive reference on the use of surface geophysical methods for groundwater investigations is given by Zohdy, Eaton, and Mabey (1974).

Land parcels that often do not lend themselves to surface geophysical methods include areas consisting of large cobbles in alluvium, areas of severely distressed geology, or areas in which (in geologic time scale) high hydraulic energy was dissipated.

Electrical resistivity. Electrical resistivity is probably the most commonly used surface geophysical method for groundwater investigations. It is economical to apply and it produces useful information.

In the direct-current resistivity method, electrodes placed into the ground transmit current through the earth, and voltage potential is measured between two points

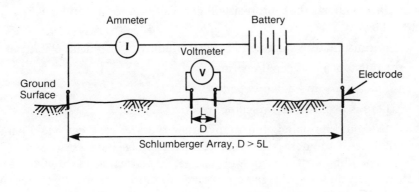

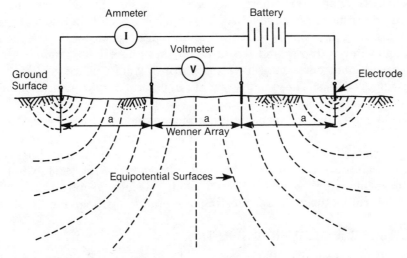

Figure 2-1 Schlumberger and Wenner electrode arrangement for measuring earth resistivity.

near the center of the generated field. Figure 2-1 schematically illustrates the most commonly used electrode arrangement—the Schlumberger and Wenner array. Sets of resistivity readings can be gathered in two ways: along a transect with constant electrode spacing (horizontal profiling) or at one location with expanding electrode spacing (electrical sounding). The first method will show apparent resistivities of materials at roughly the same depth along the transect, while the second method produces a depth profile of resistivity.

Electrical resistivity is strongly affected by water content, thus interpretations involving the unsaturated zone are quite difficult, due to the undefined distribution of moisture. In the saturated zone, resistivity is largely determined by the rock-matrix density and porosity and by the saturating-fluid salinity (electrical conductivity).

Other factors being held constant, coarse sediments with low clay content will have higher resistivity than fine-grained sediments, making possible the mapping of buried stream channels or the depth profiling of shale–sandstone sequences. Such changes in mineral quality are detectable due to the relation between dissolved solids and electrical conductivity. Resistivity values for earth materials range over more than 16 orders of magnitude, and resistivity surveys can provide useful results in most environments with simple geologic structure and distinct resistivity contrasts.

Electrical resistivity may be used in existing or new well fields, which require evaluation of suspect chemical migration from a polluting source. It is not uncommon for plumes of discharged chemicals to have relatively high dissolved-solids content.

The boundaries of such plumes may be mapped with resistivity surveys, under the proper conditions.

The Schlumberger and Wenner array (see Figure 2-1) provides for the simplest interpretation of this method. The apparent resistivity R_a is given by

$$R_a = 2\pi a \ V/I \qquad \text{(Eq 2-1)}$$

Where:

a = the electrode spacing
V = voltage
I = direct current.

The apparent resistivity characterizes a volume of earth extending below the electrode array to some effective depth of penetration. The depth of penetration is related to the electrode spacing. For the Schlumberger and Wenner array, the effective depth of penetration is commonly assumed equal to a. However, this assumption can lead to serious errors in calculated depths. The depths to horizontal boundaries are best determined by matching theoretical curves of various model conditions to the curves obtained from field measurements. The Schlumberger and Wenner array provides for better resolution of subsurface features, but is slightly more difficult to analyze. Interpretations should be performed by a trained analyst familiar with local conditions.

The resistivity method is generally limited to use in simple geologic environments, with two or three distinct layers, and where depth of penetration is limited to about 1500 ft (460 m). Best results are obtained when the depth to groundwater is small, due to the complications of unsaturated materials. Also, the method is less effective in urban areas due to the presence of buried metal pipes, wires, and similar obstructions, which dominate measurements with unwanted noise.

There are other surface geophysical methods that fall within the electrical methods category, including telluric, magneto-telluric, electromagnetic, and induced polarization methods. However, their use is not generally applicable to groundwater supply investigation. Electromagnetic methods have recently become popular for shallow groundwater investigations, especially those involving groundwater contamination. The results are similar to those derived from the direct-current resistivity method, although resolution is poorer and exploration is generally limited to the upper 180 ft (55 m). The method has received attention because it involves no direct contact with electrodes, making it quick and easy to apply in the field. The other methods mentioned have some application in specialized research, and are discussed by Zohdy, Eaton, and Mabey (1974).

Seismic refraction and reflection. Seismic methods are perhaps the most useful geophysical tools for hydrogeologic investigations, although costs are relatively high. Seismic methods use contrasts in the velocities of elastic wave propagation between different earth materials. For example, unconsolidated sands and gravels exhibit low propagation velocities, whereas crystalline rocks exhibit the highest propagation velocities. Propagation velocities are higher in saturated materials, providing for detection of the water table.

Commonly, elastic waves are initiated with the use of explosive "shots" in shallow borings, although use of truck-mounted hydraulic earth vibrators (thumpers) is becoming widespread. Lines of geophones are laid out on the ground surface to detect waves refracted or reflected from various subsurface discontinuities; this provides a measure of travel time. Travel time records can then be analyzed to produce a picture of the subsurface. As with all surface geophysical methods, the interpretation of

28 GROUNDWATER

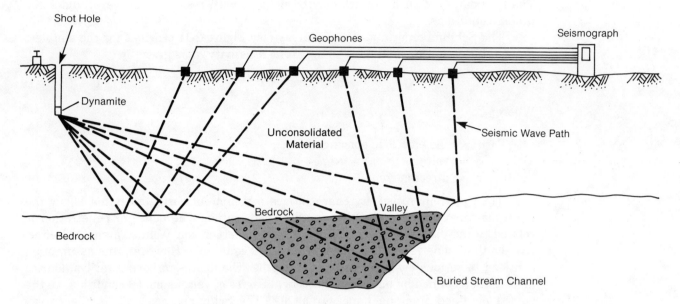

Figure 2-2 Application of seismic refraction method for reconnaissance mapping.

seismic data requires an assumed model of subsurface structure; the more preexisting information from surface geologic data and borehole logs that is available, the more reliable the results from seismic surveying will be.

There are two separate methods of seismic exploration—reflection and refraction. Seismic reflection is the method of preference for petroleum exploration. By virtue of this fact, it is the most widely used seismic method. The cost and complexity of equipment and analyses required to apply the reflection method is greater than that required for the refraction method. However, for deep exploration in multilayered environments, the reflection method is generally superior.

Seismic refraction is the only method that has been commonly applied in groundwater investigations. Besides cost considerations, this method is often suggested to have advantages in environments where deep alluvial or glacial fill exist; these are areas where hydrologic interest is high. Less preexisting knowledge is required to apply refraction seismology, and good results can usually be achieved in most groundwater investigations. The most severe limitation of seismic refraction is that return signals can only be obtained as long as each successively deep layer has a higher propagation velocity than the overlying layer. In areas of scarce water, where water wells are drilled to great depths (for example, parts of the arid southwestern United States), this limitation may prove too restrictive, and the higher cost of seismic reflection may be warranted.

Figure 2-2 shows the use of the seismic refraction method for reconnaissance mapping of the depth to bedrock and the location of a buried stream channel. Seismic refraction can be used to determine the thickness of surficial fracture zones in crystalline rock, and to map the depth and thickness of subsurface layers, up to at least the equivalent of three layers. As in the resistivity method, the analyses become very difficult and the results less reliable for more than three layers.

As long as the geologic structure is simple, depths of a few thousand feet (600–700 m) can be explored with seismic refraction, given a sufficient explosive shot (contained within a sufficiently deep shot hole). Seismic reflection can be used to gather information from over 10,000 ft (3000 m) deep. For small area applications, where only very shallow materials need to be explored, a sledgehammer struck on metal plate at ground surface might be sufficient to explore the first 1000 ft (30 m).

Other methods. There are other surface geophysical methods that can be useful in groundwater investigations, although none have been widely used due to the low benefit-to-cost ratio associated with their use. Some relatively new methods, such as ground-penetrating radar, have proven useful in groundwater contamination studies; however, the depths of penetration may be too shallow for general use in water supply applications. Gravity methods and magnetic methods have been used for general hydrogeologic investigations, although they must be viewed as supplementary methods to be applied in situations where maximum information is desired and cost is not a limiting factor. These methods are discussed by Zohdy, Eaton, and Mabey (1974).

Borehole Geophysical Logging

Borehole geophysical logging has become a standard tool in groundwater exploration. The technology of borehole logging is quite involved, and experienced specialists, as well as unique equipment, are needed to perform the logging and interpretation. The discussion here is limited in scope; the idea is to introduce the reader to the capabilities and limitations of the logging tools found to be most useful in groundwater exploration. There are many texts on borehole logging, but the reader should be cautioned that most of these texts present information tailored for use in petroleum exploration. The presence of low or nonsaline water in a formation mandates the use of special analyses for interpretation. For a detailed discussion of this subject, the reader is referred to Keys and MacCary (1971).

There are many borehole logging techniques available, although electrical, naturally occurring gamma, and caliper measurements are the most widely used.

All geophysical logs (measurements) are obtained by lowering a probe down the borehole and recording continuous measurements with depth. Logging is frequently performed during drilling operations, and quick analyses of the logs by qualified personnel provide the basis for decisions regarding well completion, including depth of casing, screened intervals, and similar points. Some types of logs must be performed in an uncased well, while other logging can be done in cased wells, thereby providing the opportunity to collect data from existing wells in an area of interest.

The use of multiple logs in a single well can be essential to provide confidence in interpretations. Each type of log measures different physical properties, and combined analysis may resolve ambiguities that would exist if the interpretation were to rely on a single log. The greater the number of wells logged in an area, the greater the statistical confidence in the data and interpretations as being representative of the subsurface environment.

It is often difficult to decide on the number of wells and the types of logs to be used in an investigation. Most groundwater investigations obtain adequate information using caliper, resistivity, spontaneous potential, natural gamma, and lithologic logging. The cost of these techniques should be evaluated in the context of the time available, accuracy needed, and the basic purpose of the survey.

Caliper logging. Caliper logging is simply a way to measure the well hole diameter with depth. A probe with levered arms is brought up through the hole, while a record of the position of the arms is made at the surface. A caliper log accounts for borehole diameter effects, which can be quite significant for use with other types of logs. The caliper log also provides indications of lithology, stratigraphic correlation, and fracture-zone locations.

Electrical resistivity logging. Electrical resistivity logging is commonly thought of as encompassing both single-point resistance logging and the more quantitative resistivity logging. However, a distinction should be drawn between these two methods. Single-point resistance (often loosely referred to as resistivity) is measured between a single downhole electrode and surface electrode, as shown in Figure 2-3. This configuration also provides for spontaneous potential logging. While single-point resistance logging is commonly used in groundwater investigations, due to lower costs, the benefits of resistivity logging may justify the additional expense. The use of point-resistance logging is generally limited to geologic correlation, determination of bed boundaries, changes in lithology, and location of fracture zones in resistive rocks.

On the other hand, resistivity logging can provide indications of formation porosity, hydraulic conductivity, and fluid resistivity (water quality), in addition to providing information on geologic correlation. Resistivity logging is performed using multiple downhole electrodes that sense the resistivity of a fairly well-defined volume of earth. There are many resistivity tools available, with various advantages and disadvantages, and the results from logging will depend on the methods selected. Consultation with trained specialists and reference to published literature (Keys and MacCary 1971) will aid in this selection.

Spontaneous potential logging. Spontaneous potential (SP) logging is inexpensive to apply, and it provides useful information regarding geologic correlation, bed-thickness determination, and separation of clay or shale layers from sand, sandstone, or carbonate layers. Spontaneous potential is the natural voltage potential

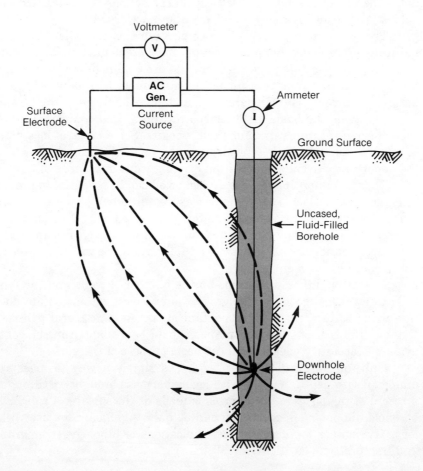

Figure 2-3 Single-point electrode arrangement for resistance and spontaneous potential logging.

that develops between differing strata and between the rock mass and the borehole fluid. The configuration shown in Figure 2-3 can be used for SP measurement by disconnecting the current source. As with the other electrical logs, SP can only be measured in liquid-filled boreholes that have not been cased.

Gamma logs. Natural gamma radiation is emitted by earth materials. A log of this emittance can be used to identify lithology and stratigraphic correlation in open or cased wells and liquid- or air-filled wells. The highest gamma counts are produced by shales and clay, so that an indication of hydraulic conductivity can be derived from gamma logs.

Gamma-gamma logs. Gamma-gamma logs are obtained by introducing a radioactive material into the borehole and measuring the intensity of back-scattered radiation. Gamma-gamma logging can be done in open or cased, liquid- or air-filled boreholes, and it provides information on lithology, bulk density, and porosity. These logs can also be used to locate cavities or cement located outside of a well casing. If measurements are made below the water table prior to pumping and in the saturated medium after pumping, it is possible to estimate specific yield from gamma-gamma logs (Davis 1967).

Neutron logging. Neutron logging is primarily used to measure water content in unsaturated materials and to measure total porosity in the saturated zone. These logs can be run in open or cased boreholes that are either liquid- or air-filled. Neutron and gamma-gamma logging tools have radioactive sources that must be registered, operated, and handled by licensed personnel.

Acoustic logging. Acoustic logging involves recording the transit time for acoustic pulses radiated from a probe in a borehole. Acoustic logging is used primarily for porosity estimation and fracture location.

Other types of logging. There are other types of logs available, such as temperature logs, fluid-conductivity logs, and fluid-movement logs, although they are generally in the realm of specialized research. One log worth mentioning is the casing log, which provides information regarding the intervals of a well that are cased, screened, or perforated. This information is essential for proper analysis of aquifer tests conducted in existing wells. Currently, water development and contamination surveys would be much cheaper and faster if logs such as these were generally available.

Log suites. Figure 2-4 depicts a qualitative interpretation of a suite of geophysical logs. Given such a complete suite of logs, quantitative information on formation porosity, hydraulic conductivity, bulk density, grain-size distribution, and fluid conductivity could be extracted. Although it is possible to analyze the data by hand, computer-aided analyses are frequently performed using digitized logs.

Aquifer Testing

Methods of aquifer testing will be discussed in detail in chapter 4. However, aquifer testing is very useful during exploration of test holes, and many effective methods are now available for performing such testing. In the simplest form, a record of flow rates of the water produced at different depths while drilling with air is a form of aquifer testing, yielding valuable information.

Exploratory Drilling

In some instances, existing wells may not be located in a potential water supply aquifer, making it necessary to drill one or more exploratory wells into the aquifer. Exploratory drilling is performed in order to determine an aquifer's characteristics, including hydraulic conductivity, water quality, thickness, and areal extent. The

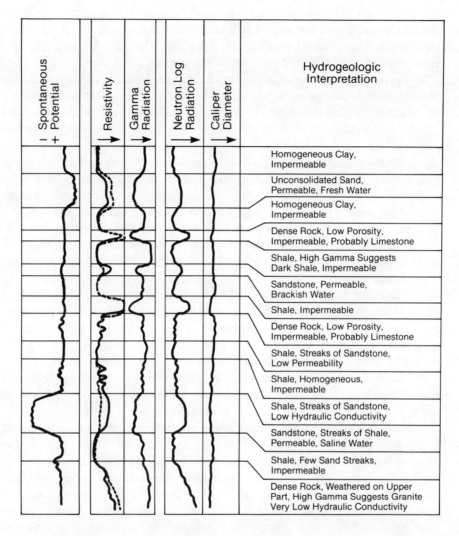

Adapted from S.N. Davis and R.J.M. DeWiest, *Hydrogeology*. John Wiley & Sons, Inc. New York [1966]).

Figure 2-4 Qualitative interpretation of a suite of geophysical logs.

newly drilled wells can also be used for borehole geophysical logging and aquifer testing as described previously. The number and spacing of the wells drilled will depend on the size of the aquifer and the amount of water supply development planned.

Lithologic Logging

The importance of accurate lithologic logging during drilling cannot be overstated. Lithologic logs and drillers' logs, including drilling rate, are used in concert with geophysical logs to gain a maximum of information regarding groundwater characteristics. Geophysical logs, which were discussed earlier in this section, can more accurately place the depth of discontinuities than a lithologic log. To obtain the best information, a geologist experienced in well logging should be employed.

Application

There is no set order in which to apply exploration methods; a balanced program applying appropriate combinations will produce the most information. However, it should always be remembered that the knowledge gained through geophysical investigation only extends the factual information that is based on test drilling and adequate sampling.

METHODS FOR MONITORING GROUNDWATER QUALITY

Prior to developing a groundwater supply, it should be demonstrated that the water quality is currently acceptable and is expected to remain so in the foreseeable future. This initial assessment can provide a basis for legal action against any parties who may contaminate the water supply after development. After the initial water quality assessment is performed and groundwater development is assured, a system for monitoring water quality should be maintained, and a reassessment of up-gradient contamination risks should be performed periodically.

Monitoring Wells

In areas where contamination risks are high, it is advisable to install sentinel monitoring wells. These wells, located at various depths, will provide definition for the initial groundwater assessment. Sentinel wells also serve as an early-warning system to detect water quality changes before they affect the water supply wells. In recent years, equipment for sampling from monitoring wells has become widely available. Small submersible or portable pumps can be installed into well casings as small as 1 in. (25 mm) in diameter.

There is no generic way to prescribe the appropriate number of monitoring wells. The number of wells necessary, locations, depths of completion, and construction details must be specified as part of an integrated plan that takes into account likely sources of contamination, local hydrogeology, and the hydraulic effects of the proposed groundwater development. What was previously considered down-gradient from the well can become up-gradient after pumping begins, and these effects should be simulated, possibly by computer-aided modeling, to aid in designing a monitoring-well network.

The initial assessment may indicate that one monitoring well is sufficient to begin with, but an increased contamination threat in future years (for example, due to local growth and development) could indicate a need for additional monitoring wells. This points out the importance of continual evaluation of changes that might affect the groundwater supply. One suggestion is to reverse the potential pollution site monitoring requirement of one well up-gradient and three down-gradient for a permitted hazardous waste site. Therefore, the water-resources-development agency would be responsible for three wells up-gradient and one down-gradient.

Sampling

It is desirable to have a multilevel sampling capability that ensures against the possibility of a monitoring well acting as a conduit for vertical migration of contaminants. Several techniques are available, including locating several wells of differing depths in close proximity to one another (cluster wells) and using multiple-completion monitoring wells, which consist of a nest of piezometers installed in a single borehole, as shown in Figure 2-5.

Proper construction of multiple-completion wells is not an easy task, but it can offer cost savings. It is critical that the construction/cure is not rushed, and care must be taken to ensure that materials are properly placed into the well bore. If completed properly, the perforated portion of each piezometer will be isolated from the others in the nest, providing for monitoring fluid pressures and water quality testing at that

34 GROUNDWATER

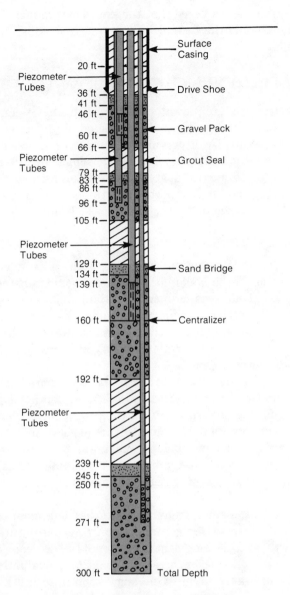

Figure 2-5 Schematic of multiple-completion monitoring well.

isolated level. Three-dimensional data collected using a network of multilevel monitoring wells will provide the only useful definition of contaminant distribution. This definition may help define origin, although the cost of such information may be high.

Analysis. Samples taken from monitoring wells should be analyzed for suspected contaminants, including severe mineralization. The mineral quality of water will limit the range of possible water uses. For example, hard water, with high concentrations of calcium and magnesium, will be unsuitable for boiler feed. Water containing high concentrations of sodium or boron will be unsuitable for irrigation. Although the biological quality of deeper groundwater is usually good, checks for fecal bacteria indicators should still be made periodically.

In recent years, a wide variety of organic chemicals, which can be harmful even in extremely low concentrations, have become a concern. Unfortunately, analyses to detect all of these chemicals can be prohibitively expensive. Thus, the analytical methods used should be directed towards detection of suspected compounds. Knowledge of probable sources of contaminant chemicals used in the area and selection of any key indicator constituents should be used in the design of the sampling and analysis program to reduce cost without loss of study credibility. Guidance for selecting chemicals to be tested may be obtained from state and federal regulatory officials responsible for facility permits. Indicator parameters, referred to as "priority pollutants," often can be used to determine the likely presence or absence of chemicals that are a concern to groundwater development.

Fortunately, groundwater quality in many locations does not change quickly, especially compared with surface water quality. Therefore, the frequency of groundwater sampling normally need not exceed quarterly or even semiannual checks, except for potable water sources or areas of suspected contamination.

FIELD LOGISTICS AND DOCUMENTATION

This section covers land costs, permits, professional services, and documentation. These points are very important to the evaluation of potential water supply sites.

Land Costs

Legal access to property for preliminary groundwater investigation will commonly be granted by the land owners on request. However, as soon as a property has been identified as attractive for detailed exploration, purchase or lease options should be obtained. More reasonable prices can be secured before confirmed knowledge of good water supplies is available. Drilling-site preparation and restoration can be added costs to the groundwater development, which should not be overlooked.

Permits

Drilling permits must be secured and fees paid (often at the county level in most states) even for exploratory drilling. This generally is the responsibility of the well driller or an engineer in charge. In addition to the driller, a person (often a qualified geologist) is commonly placed in charge of supervising the drilling and well-construction activities. This person's responsibilities will usually include procuring well-construction materials, well logging, conducting or overseeing geophysical logging, interpreting logs, well designing, and certifying as-builts.

Part of the drilling permit will include proper abandonment once the well has served its purpose. There is considerable public agency concern focused on well construction and post use of a well, due to the possibility of cross contamination between shallow zones and deeper high-quality aquifers. The documentation of all field work can be very valuable in later phases of groundwater development or protection. Proper land survey location and description of the wells and complete as-built drawings of construction are desirable.

Professional Services

Often a firm specializing in hydrology is valuable to the groundwater developer for its knowledge and experience in carrying out field work and preparing the necessary reports. Some firms offer total services in groundwater development, as well as highly specialized equipment or services. Budget constraints, complexity of the project, and adequacy of the groundwater developer's staff can determine the most appropriate mix of personnel for a successful undertaking. Some states require registered engineers, certified geologists, and other professionals to verify the accuracy and completeness of field work.

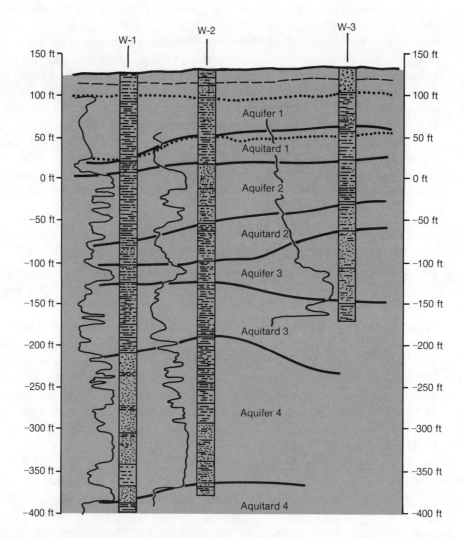

Figure 2-6 Graphic detail of hydrogeologic cross section.

Documentation

Documentation of initial investigations and water supply development must be detailed and complete. Due to the complex nature of the information, the use of graphics is helpful. Cross sections (Figure 2-6), showing the hydrogeologic interpretation, with lithologic logs, geophysical logs, and as-builts for wells, present a good summary of information. Maps showing predevelopment groundwater contours versus the contours as affected by the new pumping, including surrounding land uses and any potential sources of groundwater contamination, should be available. Figures 2-7 and 2-8 illustrate these maps.

Legal documents. Multiple copies of reports pertaining to groundwater development may be required by different levels of government for differing purposes. If the required reports are not to be submitted on issued forms, it would be useful to design a reporting system that meets the needs of all federal, state, and local organizations requiring information.

Because federal, state, and local laws are requiring more and more information to be filed on a periodic basis, permits must be filed, fees must be paid, and monitor-

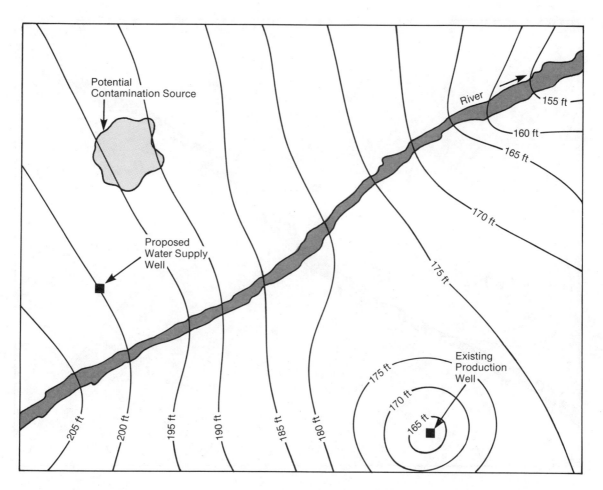

Figure 2-7 Predevelopment groundwater contours showing potential contamination source down-gradient.

ing must be conducted. These activities should be part of the planned groundwater development program. It may be desirable to consult with an attorney knowledgeable in all aspects of the law, including groundwater, land use, and permit procedures, before finalizing reporting procedures. Reports can be of great value in litigation and, therefore, should be prepared with care and be subjected to appropriate legal, technical, and managerial review.

Application. In addition to the traditional groundwater quantity reporting (along with the few inorganic analytical tests), more extensive testing for organic contaminants is being required. Water conservation interest groups are using the pumping data to predict available groundwater supply and look for indications of groundwater mining. Information is also being filed with various government agencies involved in groundwater monitoring of facilities that produce, handle, store, treat, or dispose of chemicals determined to be hazardous to health or the environment. This information, when combined with information provided by the groundwater developer, can increase understanding of the regional groundwater system under study and its potential and reliability as a water supply.

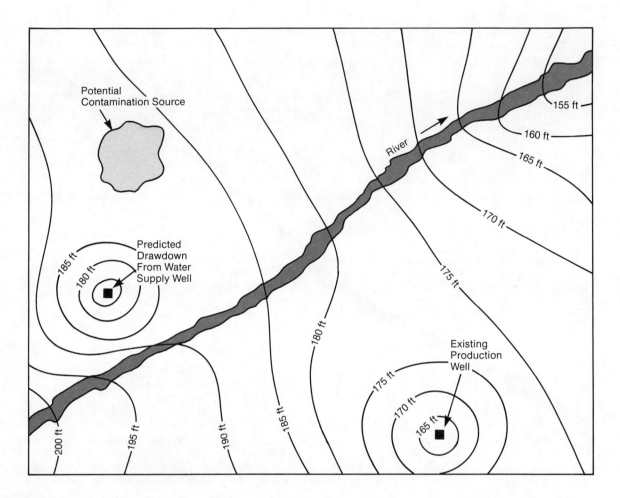

Figure 2-8 Predevelopment groundwater contours showing predicted effects of ill-advised development.

References

Davis, R.W. A Geophysical Investigation of Hydrologic Boundaries in the Tucson Basin, Pima County, AZ. Doctoral thesis. Univ. of Arizona, Tucson, Ariz. (1967).

Davis, S.N. & DeWiest, R.J.M. *Hydrogeology*. John Wiley & Sons, Inc., New York (1966).

Keys, W.S. & MacCary, L.M. Application of Borehole Geophysics to Water-Resources Investigations. US Geol. Surv. Tech. of Water Resources Invest., Book 2, Chap. E1 (1971).

Summary Appraisals of the Nation's Ground-Water Resources. US Geol. Surv. Prof. Paper No. 813, Chap. A–U (1978–1982).

Todd, D.K. *Ground-Water Resources of the United States*. Premier Press, Berkeley, Calif. (1984).

Zohdy, A.A.R.; Eaton, G.P.; & Mabey, D.R. Application of Surface Geophysics to Ground-Water Investigations. US Geol. Surv. Tech. of Water Resources Invest., Book 2, Chap. D1 (1974).

AWWA MANUAL M21

Chapter 3

Wells—Types, Construction, and Use

Such a large variety of geologic and hydrologic conditions exist today that it is impossible to construct only one type of well. Consequently, various methods of well construction have been developed, and several construction types are generally used in most areas. The reasons for selecting a particular construction type may be induced through habit, personal preference, or good salesmanship. However, selection of proper well construction should be based on thorough engineering study and design to best accommodate existing conditions or requirements.

TYPES OF WELLS AND THEIR CONSTRUCTION

A well is a hole sunk into the earth to obtain water from an aquifer. Well type generally refers to the method of well construction—dug, bored, driven, or drilled. A fifth type of well, which is not named for its type of construction, is the radial collector well. Each type of well has certain advantages based on ease of construction, storage, capacity, ability to penetrate various formations, and ease of safeguarding against contamination.

Dug Well

A dug well can furnish relatively large supplies of water from shallow sources, by using the water from all water-bearing formations that it penetrates. The yield from such a well increases with diameter, but the increased yield is not proportional to the increased size. A dug well is of large diameter, usually 8–30 ft (2–9 m) across, when installed for municipal purposes, with a depth varying from 20 to 40 ft (6 to 12 m). Because of the large opening needing protection against surface contamination, dug wells are easily polluted by surface water, airborne material, and objects falling into or finding entrance into the well.

Construction. Generally, a dug well is circular, because this shape adds strength and is usually easier to construct. Material is frequently excavated using a pick and shovel. When the depth of the well makes it impossible to throw the excavated material directly out of the hole, a hoist with a bucket is used. Clam-shell buckets with power hoists can be used when no large boulders or thick layers of clay or hardpan are encountered.

If the formation in which the well is being dug will stand without support, it may not be necessary to line the excavation until the water table is reached. When the water table is encountered, sheet piling is used to temporarily brace sides of the excavation. Later, after the placement lining or casing (usually called the curb) is placed, the piling is removed.

In order to minimize surface pollution, monolithic concrete curbs are constructed. The curbs are built in rings 3–4 ft (1–1.2 m) high. As the well depth increases and the curbs sink, additional rings are added. The rings are reinforced and are tied together by the vertical steel. Both inside and outside forms are used to get a smooth surface that will sink easily. The portion of curb that lies in the water below the limit of drawdown should be perforated. This is usually accomplished by casting short pieces of pipe in the curb, several in each square foot (0.09 m^2) of curb. Graded gravel should be placed around the outside of the curb to keep sand from coming through the perforations. Selection of pipe and gravel sizes depends on the natural formation grain size, which should be predetermined.

Bored Well

A bored well is installed where speed and economy of material are important and where water can be obtained at shallow depths by penetrating unconsolidated formations. An auger can be used only where formations, though relatively soft, will permit an open hole to be bored to depths ranging from 25 to 60 ft (8 to 18 m) without caving. The most suitable formations for bored wells are glacial till and alluvial valley deposits. Bored wells are limited to about 36 in. (1 m) in diameter. Generally speaking, this type of well has little application for municipal use.

Construction. A bored well is constructed using either hand or power earth augers. The same type of hand auger used for digging shallow holes can be used for well construction. However, extensions are needed for the auger in order to handle the greater depths encountered in well construction. Power-driven augers are of the half-cylinder, open-blade type or the cylindrical-bucket type with cutting blades at the bottom. The material cut by the blade is collected in buckets lowered into the hole and then removed.

As sand and gravel are encountered below the water table, the well casing is lowered to the bottom of the hole. Boring continues by forcing the casing down as the material is removed from the hole. After the well is completed, the annular space between the bore hole and outside of the casing should be filled with cement grout to prevent the supply from becoming contaminated.

Driven Well

Driven wells are practical only where the water table is near the land surface. They are simple to install and economical in terms of time and labor. A driven well consists of a pointed screen, called a drive point, and lengths of pipe attached to the top of the drive point. The drive point is a perforated pipe covered with woven-wire mesh, a tubular brass jacket, or is similar to screens for drilled wells and is adaptable to driving. At the base of the drive point is a pointed steel tip for breaking through pebbles and thin layers of hard material and for opening a passageway for the point. This type of well varies from $1\frac{1}{4}$ to 4 in. (32 to 100 mm) in diameter and is a maximum of about 30–40 ft (9–12 m) deep.

In municipal practice, the driven well is used where thin deposits of sand and gravel are found at shallow depths and the production rate of the formation is limited. Under such conditions, a single well could not produce a sufficient quantity of water; however, a battery of well points, with the wells located a reasonable distance apart and connected by a common header to the pump, might develop sufficient water to supply a small community. In this case, a suction-type pump would have to be used in which the water would rise in the wells to a point where the pump could pick it up, preferably about 15 ft (70 m) from the land surface. Driven wells are also used as observation wells during aquifer tests.

Construction. It is considered good practice, when constructing a driven well, to first install an outer casing. This protects the inner casing to which the pump is attached and on which a partial vacuum is placed that can draw contamination at leaking joints. This outer casing is usually 2 in. (50 mm) larger in diameter than the well casing. It should extend a minimum distance of 10 ft (3 m) below the ground surface.

In sand and gravel, it is preferable to extend the outer casing to a point just above that at which the drive point is to be set. The outer casing can be driven by using a sledgehammer or a tripod and pulley, which raises and lowers a heavy block onto a drive cap placed on top of the casing. Extra-heavy pipe should be used in order to withstand the load placed on it when driving. The sand and gravel in the outside casing are removed by an auger as driving proceeds. If clay is being penetrated and difficulty would be encountered in driving, the outside casing should be set in a hole prepared with an auger. Under such conditions, a 10-ft (3-m) depth usually affords sufficient protection. After the casing is set, the annular space between the bore hole and the outside of the casing should be sealed with cement grout.

The next step is to lower the drive point, which is attached to the bottom of a string of inner casing, into the hole. The drive point is driven below the bottom of the outside casing to the depth where it is expected that a water-bearing formation is located. This depth may be tested periodically by pouring water down the well casing and observing how quickly the water moves down the well, by attaching a pump, or by jetting water down inside the casing to wash out the screen. When the final depth has been reached, raising or lowering the pipe 1 ft (0.3 m) or so frequently brings a greater portion of the screen into contact with the water-bearing formation, resulting in increased production.

Drilled Well

A drilled well is commonly used for municipal purposes. Its depth is limited only by the distance one must drill to obtain an adequate supply of fresh water. There is a wide range in pipe diameter— from 2 to 48 in. (50 to 1210 mm) and above—for drilled gravel-walled-type wells. Likewise, high-capacity wells can be developed that produce up to thousands of gallons per minute, provided, of course, that these quantities are present in the aquifer.

Construction. Drilled wells are very versatile and able to develop water from shallow or deep sources in unconsolidated sands and gravels or rock. When constructing the well, a drilling rig is used to excavate or drill a hole, and then a casing is forced down the hole to prevent it from collapsing. When a water-bearing formation of sufficient capacity is found, a screen is set in place, permitting water to flow into the casing and holding back the fine material. When the drilled well passes through rock, no screen is used.

Wells are drilled using many different types of equipment. The construction methods can be classified as cable-tool, rotary, reverse-circulation rotary, California, and jetting. Each method will be briefly described.

Cable-tool. The cable-tool method of drilling is used extensively for wells of all sizes and depths. It is referred to by various names, including percussion, spudder, and solid tool. There is an infinite variety in the details of construction and operation of the drilling machines. Yet, all machines are used to perform the same function, that is, to dig a hole using the percussion and cutting action of a drill bit that is placed on the bottom of a string of solid drilling tools. The drilling tools are placed at the end of a cable that is alternately raised and dropped by suitable machinery. The drill bit, a club-like, chisel-edged tool, breaks the formation into small fragments, and the reciprocating motion of the drilling tools mixes the loosened material into a sludge.

Generally, several feet (one to two metres) of hole are drilled at each run of the drill tools. After each run, the tools are pulled from the hole and swung aside while a bailer is used to remove the sludge. The bailer consists of a 10- to 25-ft (3- to 8-m) long section of tubing with a check valve in the bottom. The bailer is smaller in diameter than the drill hole so that it can move up and down freely. Alternate drilling and bailing is continued.

An experienced driller is needed to adjust the length of the drill cable so that the bit will strike with the right amount of weight and stroke. The driller holds the cable and notes the character of the jar when the cable is dropped, which indicates the manner in which the tools are operating. The driller regulates the length of stroke and rapidity of blows according to interpretation of the cable vibrations.

Drill hole characteristics. It is important that the drill hole be straight and plumb. Usually the first indication that the hole is out of plumb is that the drilling tools begin to stick. When sticking occurs, drilling should stop and the hole brought back to straight and vertical. A straight, vertical hole permits the lowering of a pump to the desired depth and prevents damage to pumping equipment. Pump manufacturers state that pumps will operate satisfactorily when slightly inclined. However, a well out of alignment and with bends or curves causes severe wear on the pump shaft, bearings, and pump column pipe. In extreme cases, it may be impossible to lower a pump into this type of well.

Generally, reasonable care on the well driller's part is all that is required to drill a well that is straight and plumb. The drilling contractor should be required to demonstrate, through testing, that the completed hole is straight and plumb. In one such test, a 40-ft (12-m) length of pipe is lowered into the well to the depth of the lowest anticipated pump setting. The outer diameter of the test cylinder should not be more than $1/2$ in. (13 mm) smaller than the diameter of the casing being tested. The test cylinder should move freely throughout the length of the casing to establish the alignment of the hole. As to plumbness, the well should not vary from the vertical by more than two thirds the inside diameter of the well per 100 ft (30 m).

Formation recognition. During cable-tool drilling, formation recognition is limited to the hardness of the formation being drilled. This, in turn, is distinguished by the distinct vibration of the various materials. To determine the exact kind of material being drilled, it is necessary to take samples of the cuttings at 5-ft (1.5-m) intervals or at each noticeable change in formation. Generally, it is not difficult to detect water-bearing formations. A sudden rise or fall of the water level in the well frequently indicates that a permeable formation has been encountered. Sand, gravel, sandstone, and limestone formations produce the largest quantities of water. It is advisable, therefore, to be especially watchful when drilling in these formations.

Advantages. With the cable-tool drilling method, a more accurate sample of the formation is obtained; it is possible to test quantity and quality of each formation as drilling proceeds; less water is necessary for drilling operations; and, in most cases, a cable-tool rig is light and can traverse rough country easily.

Rotary. As the name implies, in the rotary method of well drilling, the hole is made by rapid rotation of a bit on the bottom of a string of drill pipe. The speed of rotation of the drill pipe and the bit can be varied according to the type, form, and size of bit; the formation to be drilled; and the strength of the drill pipe.

The drill pipe is hollow, allowing fluid to be pumped down inside it to the bit. When drilling commences in clay soil or mud, drilling fluid of sufficient viscosity, which helps lift cuttings to the surface, and with necessary sealing qualities and weight, which help stabilize the well bore, may be formed by normal drilling operations, starting with clear water. If, on the other hand, the formation near the surface is sandy, it is necessary to mix the mud or drilling fluid before starting. This fluid is prepared in a pit located near the drilling rig. Native clays are sometimes used, but commercial colloidal material (bentonite clay base), which is purchased in powdered form and mixed with water, is preferable. A recently developed and patented material, which has an organic rather than a mineral base, appears to have many superior characteristics, one of which is its ability to break down after three or four days. Thus, no hard-to-remove mud cake is formed on the walls of the bore hole.

The prepared fluid circulates through the drill pipe and out through holes in the bit, where it sweeps under the bit, picks up the material loosened by it, and carries it to the surface. The fluid from the well overflows into a ditch and passes into a settling pit, where the cuttings settle out. The fluid, now free from coarse materials, flows into another pit, where it is picked up by the pump for recirculation.

Reverse-circulation rotary. The reverse-circulation rotary method differs from the straight rotary method in that the fluid circulates in the opposite direction. The pit is constructed so that the drilling fluid will flow down the bore hole and rise in the drill pipe, carrying the cuttings with it. A high-capacity (500 gpm [1900 L/min] or greater) pump is attached to the drill pipe and keeps the fluid moving at high velocity. The pump may discharge to waste if a large fresh supply of water is available, or the cuttings may be permitted to settle and the fluid recirculated.

The reverse-circulation method primarily uses clear water with no mud additives. Keeping the bore hole open requires a large volume of water to maintain a head well above the natural static water level. The head results in a flow into the formation from the bore, rather than the reverse, and thus prevents the wall from caving. When substantial thicknesses of material that will not accept water are encountered, caving may result from the wash action of the fluid moving down the hole. If a highly permeable formation is encountered and the water supply at the surface is limited, it may be necessary to "mud the hole" by adding processed clays, which will result in a plastered hole similar to the type of hole produced by the straight rotary process. Needless to say, any mudding defeats one of the primary advantages of the reverse-circulation method of construction. The reverse method is particularly well suited for artificial-gravel-pack wells, because mud-cake formation on the face of the bore is minimized and development time is reduced considerably.

The actual boring or cutting into formation, with either the rotary or reverse-circulation rotary method, is accomplished by the bit at the lower end of the drill pipe. There are several types of bits, and the choice of a particular bit depends on the preference of the driller and the nature of the formations to be drilled. Initial pressure is applied on the bit by the weight of the column of drill rods, aided by the hydraulic feed. The rotating bit disintegrates the formation by shearing and cutting action. It must not cut the formation faster than the circulating fluid can carry the cuttings away.

Advantages. Both rotary methods can drill large holes up to 5 ft (1.5 m) in diameter. A second advantage is that a test well can be drilled and, if desired, the hole can be abandoned at a minimum expense by plugging. Plugging can be performed without the trouble of pulling the casing or leaving several strings of casing in the hole. In soft, loose, unconsolidated materials, such as dune sand and quicksand, the hole may be difficult to keep open unless the rotary method is used. It also is generally faster than cable-tool drilling.

California. The stovepipe, or California, method of well construction was developed in California primarily for sinking water wells in unconsolidated alluvial materials that consisted of alternate strata of clay, sand, and gravel. Wells 16–20 in. (400–500 mm) in diameter and up to 300 ft (90 m) in depth are constructed using this method. The California drilling method uses the same general principles used in the standard cable-tool method. However, in the California method, a specially designed bucket is used as both bit and bailer, and short lengths of sheet metal, either riveted or welded together, are used for casing.

The mud-scow bit used in this method consists of a disk-valve bailer with a sharp-edged cutting shoe on the bottom. It is similar to an ordinary sand bucket, except that it is heavier, larger in diameter, and has the cutting shoe on the bottom. Each time the bit is dropped, some part of the cuttings are trapped in the bailer, which, when filled, is pulled to the surface and emptied.

At the bottom of a string of California stovepipe casing is a riveted steel starter, 10–25 ft (3–8 m) long, made of three thicknesses of sheet steel with a forged-steel shoe at the lower end. This reinforcement prevents the bottom from collapsing when pressed down under pressure. Above the starter, the casing consists of two sizes of sheet steel made into riveted or welded lengths from 2 to 6 ft (0.6 to 2 m). The larger-size casing is made to fit snugly over the smaller size. Each outside section overlaps the inside section by half its length so that a smooth surface results both outside and inside when the casing is in place. In this way, the inner and outer joints never coincide.

The casing is forced down, length by length, by hydraulic jacks anchored to two timbers buried in the ground. These jacks press on a suitable head attached to the upper section of the stovepipe casing so that the end of the casing will not be telescoped. The casing may also be driven by raising and lowering the tools with a driving head.

After the casing is down, it is perforated in place using a Mills knife or similar device. This perforation is made by tearing the metal. Care must be exercised not to make the openings too large and not to perforate too much area.

Jetting. The final method of drilled well construction, jetting, is used to drill a vertical well. This method is particularly successful when water is found in sand at shallow depths; however, it is also used for deep wells.

Jetting equipment consists of a drill pipe, or jetting pipe, which is equipped with a cutting bit on the bottom. Water is pumped into the well through the drill pipe and out of the drill bit against the bottom of the drill hole.

Casing usually is sunk as drilling proceeds. In some instances, the casing will sink a considerable distance under its own weight. Ordinarily, however, a tripod and drive weight are needed to force it in place. As a rule, one size of casing is used for the entire depth of the well. However, if a well is rather deep, it is difficult to drive a single string of casing using this method, and an additional string of smaller casing is used.

After the casing is lowered to the water-bearing formation, the well screen and pipe are lowered into the casing. The outside casing provides protection to the inner casing connected to the screen. The well screen is exposed to the water-bearing formation by pulling back the outer casing a distance equal to the length of the screen.

Certain conditions can make this method of well construction difficult to perform. Rock formations and boulders are barriers that cannot be overcome. Formations of clay and hardpan are other types of materials that ordinarily present problems.

Radial Well

The radial, or horizontal, well has recently come into wide use. Carefully designed and constructed, this type of well can produce very large quantities of water. Generally, it is located along the shore of a lake or river to take advantage of infiltration from the water body.

A radial well can be thought of as a combination dug well and a series of horizontal driven wells projecting outward from the bottom of its vertical walls. The main well, or central caisson, serves as a collector for the water produced from the horizontal wells. The horizontal wells must be located in good, coarse, natural material, since it is somewhat difficult to accomplish any major development around the screen sections.

Water from this type of well may require a higher degree of treatment than water from other vertical wells simply because of its proximity to a surface source and thus its dependence on the quality of that surface supply. Consequently, consideration should be given not only to bacterial quality, but also to turbidity, color, toxicity, tastes, and odors.

Construction. The central caisson of a radial well is constructed of reinforced concrete. It has an outside diameter of 15–20 ft (5–6 m). The wall is generally 12–18 in. (305–460 mm) thick and is poured in circular sections 8–10 ft (2–3 m) high. The bottom of the first section, or ring, is formed with a cutting edge to facilitate the caisson's settling in the excavation.

Material is excavated from within the caisson, and care is exercised to ensure that the caisson is as plumb as possible. Each section is keyed and tied to the previously poured unit to achieve structural stability and watertightness. The final depth of the cutting edge is several feet (one metre) below the bottom of the water-bearing formation. Following the final positioning of the caisson, a concrete plug is poured in the bottom.

Laterals project horizontally through wall sleeves from the central caisson into the water-bearing formation. The laterals, which may be constructed of slotted or perforated pipe or may be conventional well screens, are positioned $1\frac{1}{2}$–2 ft (0.5–0.6 m) above the bottom of the water-bearing formation. The entire length of a lateral is perforated, with the exception of approximately a 5-ft (1.5-m) blank extending from the outer wall of the caisson. A gate valve is installed on each lateral inside the caisson to make it possible to cut off the flow into the caisson for dewatering. A superstructure is erected on top of the caisson for housing the pumps, piping, and electrical controls.

Gravel-Wall Well

A gravel-wall well is installed to permit the use of larger slot sizes in the well screen section than would be possible if the area surrounding the screen were not gravel-treated. The amount of open area in the screen is increased, as is the effective diameter of the well. With proper construction, entrance of fine sand from the water-bearing formation into the well is controlled. The benefits of this type of construction are lower entrance velocity at the screen openings and increased flow per unit of head loss.

It should be emphasized that a gravel-wall well must be thoughtfully designed. The material used in the gravel treatment must be clean, washed gravel composed of well-rounded particles. Their size will depend on the size of the natural formation. Without proper gravel size, the fine sand will not be bridged out and the yield will be adversely affected. At the same time, the uniformity coefficient* of the gravel treatment should be 2.0 or less. The size of the gravel (individual grains) should be four to five times larger than the median size of the natural material. The slot size for the screen is chosen to retain 90 percent of the pack material.

Construction. In the construction of gravel-wall wells, also commonly referred to as gravel-packed wells, selected gravel is placed between the outside of the well screen and the face of the water-bearing formation. This method of well construction is especially useful when developing water from formations composed of fine material of uniform grain size. A gravel-wall well is actually a large-diameter drilled well, but instead of the natural formation being developed and the fines being drawn into the well, leaving the coarse material packed naturally around the outside face of the screen, coarse material is artificially placed around the screen.

Gravel envelope. The most common construction used is the gravel envelope (Figure 3-1), where the outside casing is first sunk to the bottom of the formation. The diameter of this outer casing may vary from 18 to 72 in. (460 to 1800 mm). Experience and research have shown that artificial gravel packs between 6 and 8 in. (150 and 200 mm) thick perform best and require the least maintenance. For example, a 24-in. (600-mm) bore would be finished with a 10- or 12-in. (250- or 300-mm) diameter screen and 6- to 7-in. (150- to 180-mm) thick pack.

After the outer casing is in place, the screen is lowered to the bottom of the well and centered. Selected gravel is added to the annular space between the screen and the casing through a small-diameter tremie pipe. The gravel is placed evenly around the screen in 2- to 4-ft (0.6- to 1.2-m) layers. As the gravel is added, the casing and tremie are slowly raised, and the procedure continues until the entire screen is surrounded with gravel and the pack extends several feet ($1/2$ to 1 m) above the top of the screen. The outer casing is pulled back high enough to expose the entire screen section. As a rule, the screen is attached to an inner casing, extending to the land surface, into which the pump is placed. About 25 ft (8 m) of the outer casing may be left in position. If it is all removed, the gravel treatment must not extend to near the land surface, and the annular space between the working casing and undisturbed earth must be sealed with cement grout or puddled clay to prevent contamination from seeping down into the formation.

Other construction. Other types of construction include underreamed and bail-down methods. The underreamed gravel-wall well is drilled to the top of the water-bearing formation and the casing set. The formation is then underreamed, and the gravel and screen section are placed. The bail-down method makes use of a cone-bottomed screen. The outer casing is sunk to the top of the water-bearing formation and the screen bailed into place while gravel is fed into the space between the inner and outer casing. In another method, the so-called pilot-hole method, the gravel is fed into the formation through small pilot wells evenly spaced around a central well as the fines are withdrawn through the central well.

---

*The uniformity coefficient is the 40 percent retained grain size in a sieve analysis of sediment divided by the 90 percent retained grain size.

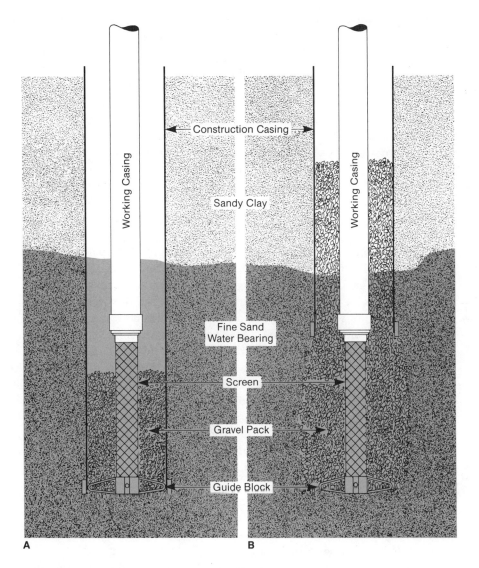

A—gravel-wall well with casing in place; B—completed gravel-wall well.

Figure 3-1 Two phases of gravel-wall well construction—gravel envelope method.

COMMON WELL COMPONENTS

Well components that are common to most wells include well casings, cementing or grouting of wells, and well screens. These components, as they relate to well construction, will be discussed.

Well Casings

The well casing is simply a lining for the drilled hole and, as such, maintains the open hole from the land surface to the water-bearing formation. It also seals out contaminated water originating from the land surface and undesirable water from formations above the aquifer in which the well is developed. For the casing to be entirely effective, it must be constructed of suitable materials and be properly installed so as to be watertight for its entire depth.

Materials. Materials commonly used for well casings include wrought iron, alloyed or unalloyed steel, and ingot iron. In recent years, well casings manufactured from fiberglass and polyvinyl chloride (PVC) have been used in certain installations. In selecting a suitable material, one must consider the strain to which it will be subjected during installation and the corrosiveness of the water and soil with which it will come in contact. Experience has shown that wrought iron and steel give satisfactory service, with wrought iron being preferred for protection against corrosion. Ingot iron is frequently used in constructing gravel-wall wells or other large-diameter wells.

Many weights of steel and wrought-iron casings are available, and it is not sufficient merely to order a certain nominal diameter of casing. The desired weight per foot of pipe must be specified. The tables in AWWA Standard A100-84 present data on steel and wrought-iron pipe recommended for use as permanent well casings.

Lighter-weight materials may be used for test wells or temporary casings. Temporary casings are sometimes used as forms when a grout seal is placed around the outside of the permanent casing; the temporary unit is withdrawn as the grout seal is placed. Under such circumstances, lighter and less-expensive material is warranted.

Joints for permanent casings should have threaded couplings or should be welded to ensure watertightness from the bottom of the casing to a point above grade. This precaution will prevent the entrance of surface contamination and undesirable underground waters above the water-bearing formation.

Installation. When drilling a well by the cable-tool method, the casing is driven as soon as it becomes necessary to prevent the ground formation from caving. A drive shoe, attached to the lower end of the pipe, keeps the hole from collapsing. Drive shoes are threaded or machined to fit the pipe or casing, and the inside shoulder of the shoe butts against the end of the pipe. Drive shoes are forged of high-carbon steel, without welds, and are hardened at the cutting edge to withstand hard driving.

Casing is driven using drilling tools, drive clamps, and the drive head. A length of casing is attached to the previous length already set. A drive head is attached to the upper end of the casing to protect it from the driving blows of the drive clamp, which is attached to the drill stem. When the drill is lowered into the length of casing and subsequently raised and lowered, the action of the dropping clamp on the drive head forces the casing into the drill hole.

Wells constructed using rotary methods are not usually cased until drilling is completed. Since the casing is smaller than the drilled hole, no driving is required.

Special situations. If the formation being penetrated is of a caving nature for the full depth of the well, a single casing is usually sufficient. In these situations, the sand and gravel caves in around the outside of the casing and closes the space between the drill hole and the casing.

If additional protection is desired against corrosion and pollution, it may be provided by installing an outer casing and filling the annular space between the casings with cement grout. With this type of installation, the outer casing may be either left in place or withdrawn completely. If withdrawn, the grout is placed as the temporary casing is removed. The temporary casing is generally several inches (50–60 mm) larger in diameter than the outside diameter of the couplings of the protective casing. This type of grouted installation may also be used where the water-bearing formation underlies clay, hardpan, or other noncaving formations.

Where the well penetrates water-bearing rock underlying unconsolidated material, it is common practice to simply drive the casing into the rock and attempt to obtain a good seal. Unfortunately, there is no certainty of obtaining a tight seal that will prevent pollution or the unconsolidated material above from entering the well. Accordingly, one way to obtain additional protection is to drive the casing down to stable rock. The rock is then drilled and underreamed to a diameter 2 in. (50 mm) larger than the outside diameter of the shoe and for a depth of 10 ft (3 m). The underreamed portion of the drill hole below the bottom of the casing is then filled with cement grout, and the casing is driven to the bottom of the hole. Before drilling is resumed, the cement grout is allowed to set up for several days, which provides a good seal from above. Drilling is continued and the cement grout drilled out. An open, uncased hole is then constructed in the water-bearing rock below this point.

Formations, such as limestone, that are channeled or creviced frequently yield polluted water. Consequently, these formations should be avoided as a supply source, unless overlain with an adequate thickness of unconsolidated formations over a rather extensive area. Under such circumstances, reasonable protection is afforded if watertight construction of the well is provided to a depth greater than that of the deepest existing well of questionable construction in the area and substantially below the lowest anticipated water level in the well. The watertight construction is achieved by drilling the hole in the creviced rock 2 in. (50 mm) larger than the outside diameter of the casing couplings and filling the annular space between the drill hole and the outside of the casing with cement grout. In some areas, such construction may not be realistic, because available water is cased off, and other methods of assuring adequate water quality may be necessary if a creviced formation is to be used.

Cementing

Water wells are cemented, or grouted, and sealed for the following reasons:
- to protect the water supply against pollution,
- to seal out water of an unsatisfactory chemical quality,
- to increase the life of the well by protecting the casing against exterior corrosion, and
- to stabilize soil or rock formations of a caving nature.

When a well is drilled, an annular space surrounding the casing is normally, and sometimes purposely, produced. Unless this space is sealed, a channel exists for the downward movement of water. In caving formations such as sand, the opening would tend to be self-sealing. In clay or other stable formations, this space must be sealed in order to prevent entrance of contamination from the land surface or creviced formations connected with the surface. Cementing minimizes this possibility of contamination.

Protection of the casing exterior against corrosion is provided by encasing it in cement grout, as described earlier in the section on casing installation. Usually about a 2-in. (50-mm) thickness of grout is recommended.

When formations located below the depth of the protective casing are known to yield water of an unsatisfactory chemical quality, such formations may be sealed off with liners set in cement grout for their entire length.

When a casing is extended to a consolidated formation lying below an unconsolidated formation, the best way to prevent sand or silt from entering the well at the bottom of the casing is by underreaming and cementing.

Materials. Materials used for cementing wells should be of a consistency that will facilitate proper placement and, thereafter, assume a permanent and durable form. Portland-cement grout, properly prepared and handled, meets these requirements adequately.

Proper preparation of the grout mixture is of utmost importance. Best results are obtained from neat cement and water mixed in the ratio of one bag of cement to not more than $5^{1}/_{2}$ gal (20 L) of clean water. Under certain conditions, other materials may be used to accelerate or retard the time of setting, to lubricate the grout mixture, and to provide roughage for sealing large crevices. Regardless of the materials used, cement, additives, and water must be mixed thoroughly.

The grout must be applied in one continuous operation to assure a satisfactory seal and be entirely in place before the initial set occurs. It is also essential that the grout always be introduced at the bottom of the space to be grouted to avoid segregation of materials, inclusion of foreign materials, or bridging of the grout mixture.

Methods. Various methods are used for placing grout, including the dump-bailer method, air- or water-pressure drive, and pumping. Other methods of grouting, not discussed here, are used by well-cementing companies, which hold patents on their procedures.

The dump-bailer method is perhaps the simplest. The cement grout is lowered in a dump bailer that discharges its load when it reaches the bottom of the hole. After the necessary amount of grout is placed in the well, the casing is pulled up far enough so that the shoe is above the grout. A plug is placed in the bottom of the casing, which is then driven to the bottom of the hole, displacing the grout into the annular space around the outside of the casing.

If the annular space outside the casing is of sufficient width ($1\frac{1}{2}$–2 in. [40–50 mm]) to accommodate a grout (tremie) pipe, air- or water-pressure drive is quite satisfactory. The tremie pipe should extend to the bottom of the annular space. Grout is then pumped into the pipe. The pipe should remain submerged in grout during the entire time the grout is being placed. The pipe may be left in place or gradually removed.

The pumping method of grout application involves installation of a pipe within the casing. The casing is suspended slightly above the bottom of the drill hole, and a suitable packer connection is provided at the bottom of the casing. The packer permits removal of the grout pipe and prevents grout leakage into the interior of the casing. Grout is pumped into the pipe and forced upward into the annular space. When the space is filled, the grout pipe is removed. Work on the well is not resumed for at least 72 h, after which time the packer connection and plug are drilled out.

Screens

Generally, wells completed in unconsolidated formations, such as sands and gravels, are equipped with screens. Screens allow the maximum amount of water from the aquifer to enter the well with a minimum of resistance and prevent the passage of sand into the well during pumping.

Although one purpose of a screen is to prevent sand from entering the well during pumping, a screen actually permits the entrance of fine-formation particles into the well during the development process so they may be removed by bailing. At the same time, the large particles of sand are held back, forming a graded natural-gravel screen around the well screen itself. In this way, the hydraulic conductivity of the water-bearing formation around the well screen is greatly increased, resulting in lower velocity head loss and higher capacity per foot of drawdown.

Selection. The selection of the proper screen is extremely important in the design of a well drawing on unconsolidated aquifers. Selection is often a complicated matter that demands a highly specialized knowledge of well construction and operation, and it is usually advisable to consult a reliable screen manufacturer. Most manufacturers maintain a screen-selection service and will make mechanical analyses of samples and recommend the proper opening.

Slot size. The size of screen openings, or the slot number, is usually expressed in thousandths of an inch (Figure 3-2). The width of the slot, or slot size as it is

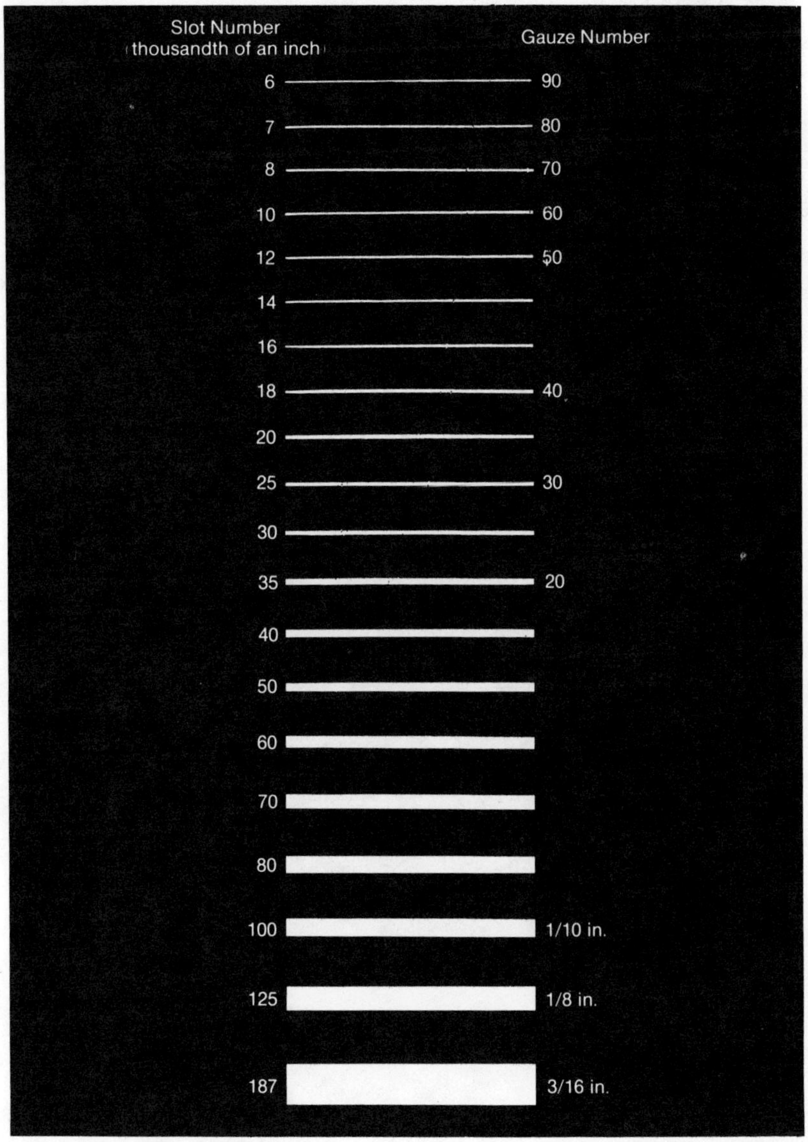

Figure 3-2 Scale of screen-opening sizes.

normally referred to, is best determined on the basis of a mechanical sieve analysis of a sample of the water-bearing formation.

As indicated, the slot size is selected so as to permit a percentage of the formation material to pass through it. This amount usually ranges between 35 percent and 65 percent, depending on uniformity of the material and the overlying formation. Therefore, it is very important that representative samples of the formation be selected for use in mechanical grain-size analysis. A complete descriptive log of the well should be submitted to the screen manufacturer with the samples. Information concerning the well diameter, aquifer thickness, transmissivity or hydraulic conductivity, and the desired well capacity should be included with the log.

It is sometimes advisable to construct an artificial gravel-pack well to permit an increase in screen slot size. When a well screen is surrounded by an artificial gravel

52 GROUNDWATER

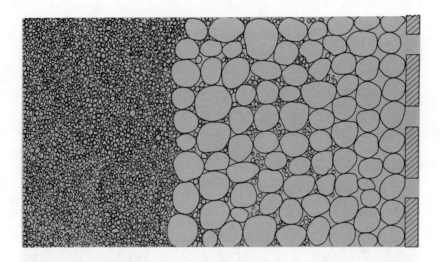
Gravel-wall well

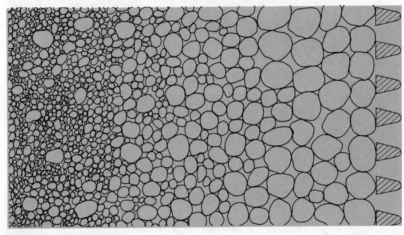
Properly developed drilled well in natural formation

Figure 3-3 Cross-sectional comparison of well walls.

wall, the size of the openings is controlled by the size of gravel used and by the types of openings (Figure 3-3).

Velocity. It is important to keep water entrance velocity through the screen openings between 0.1 and 0.2 ft/s (0.03 and 0.06 m/s). Such velocities will minimize head losses and chemical precipitation. Screen entrance velocities are computed by

$$V = Q/A \qquad \text{(Eq 3-1)}$$

Where:

V = velocity, in ft/s
Q = well capacity, in ft^3/s (1 ft^3/s = 449 gpm)
A = effective area of screen, in ft^2.

The effective screen area must be estimated carefully. It is not uncommon to allow for as much as 50 percent plugging of the screen slots by formation particles. The total open area required must be determined by adjusting either the length or diameter of the screen, because the slot is not arbitrary. The balancing of other screen dimensions to obtain open area will be discussed later in this manual.

Materials. Well screens are available in a wide range of materials, including plastic, mild steel, red brass, bronze, and stainless steel. Selection of a suitable material requires knowledge of soil and water corrosivity, intended use of the well, and anticipated cleaning or redevelopment methods. AWWA Standard A100-84 gives specific information and specifications for screen selection.

SANITARY PROTECTION

All water supply wells must be provided with adequate sanitary protection. Keys to sanitary protection are proper construction and disinfection.

Sanitary Construction

It is important that wells be developed from formations sufficiently deep to be protected from surface contamination. The minimum depth at which safe water can be obtained will vary with different soil formations and surrounding conditions. Experience has indicated that in unconsolidated materials water obtained from depths of 25–30 ft (8–9 m) or more is reasonably well protected. In other words, the well casing should extend at least to that depth, and the screen should be below it.

If it is necessary to develop a well at a depth less than that recommended where the ground formation is pervious material, some protection can be afforded by means of an impervious layer of soil at the land surface. A layer of well-compacted clay at least 2 ft (0.6 m) deep should be placed on the land surface for a radius of 50 ft (15 m) around the well. The clay layer will minimize percolation from surface water and tend to divert it to the edge of the clay and away from the well.

Another means of protection is to submerge the well screen below the pumping level of water in the well. The water level should never be drawn down into the screen section, and pump capacities should be selected to ensure from 5 to 10 ft (1.5 to 3 m) of water over the top of the screen at maximum drawdown.

A final method of protection, which was discussed earlier, is cementing. Normally, drilling a well produces annular space surrounding the outside of the casing. In caving formations such as sand, this opening would tend to close and be self-sealing. In clay, shale, and rock, unless the space between the outside of the casing and the face of the drill hole is sealed, a definite channel is provided for the downward movement of surface contamination or shallow groundwater to the water-bearing formation. Consequently, the space must be sealed securely with proper grouting. A concrete platform should then be constructed around the well.

Disinfection

During the process of well construction, the drill hole is subject to contamination from the land surface. Contamination can also be introduced by tools; drilling mud, in the case of the rotary method; the casing; and the screen. Normally, extended pumping would rid the well of this contamination; however, disinfecting the well with chlorine will accomplish the job much more quickly.

Many disinfection methods are available, and the one selected should be specified by the engineer supervising the installation. As a general rule, sufficient chlorine must be thoroughly mixed with the water in the well casing to produce a concentration of at least 50 mg/L. This solution must come in contact with the pump and discharge piping. Disinfection is achieved by adding chlorine in the casing and producing a mix by alternately starting and stopping the pump or by other methods.

The material used for gravel treatment, even though washed and clean, still carries contamination. Therefore, gravel-wall wells are sometimes difficult to disinfect following construction. In addition to the procedures outlined for disinfection, it is good practice to occasionally add tablet or powdered calcium hypochlorite by hand to the gravel-filling tube as the gravel is placed.

Even with disinfection, the water pumped from a well may still show evidence of contamination. Under such circumstances, about the only solution is to install a chlorinator at the well and treat all the water discharged to the system. In time (perhaps as long as three or four months), normal pumping will rid the well of contamination; however, during this period, chlorination to a free chlorine residual will make it possible to use the water. Additional information on disinfection is available in AWWA Standard A100-84.

Reference

Standard for Water Wells. AWWA Standard A100-84. AWWA, Denver, Colo. (1984).

AWWA MANUAL M21

Chapter **4**

Quantitative Evaluation of Wells

After a qualitative areal evaluation has confirmed the presence of water-bearing materials, it is necessary to determine how much water may be withdrawn from a well. The basic characteristics of an aquifer that must be quantitatively evaluated are the transmissivity and storage coefficient. These characteristics, discussed in detail in chapter 1, are essential to well evaluation.

TRANSMISSIVITY AND STORAGE COEFFICIENT

Transmissivity T is related to the hydraulic conductivity of the aquifer and is the rate at which water flows through a vertical strip of aquifer 1-ft (0.3-m) wide and extending through the full saturated thickness, under a unit hydraulic gradient. Transmissivity is expressed in gallons per day per foot (gpd/ft) (square metres per day [m^2/d]). Values range from less than 1000 to more than 1 million gpd/ft (less than 12 to more than 12,000 m^2/d); values of 10,000 gpd/ft (120 m^2/d) or more can be adequate for municipal and other large requirements.

The storage coefficient S represents the volume of water released from or taken into storage per unit of aquifer storage area per unit change in head. Values of S are expressed as a decimal. For unconfined aquifers, they generally are in the range of 0.3–0.01; for confined aquifers, most values are in the range of 0.005 to 0.00005.

Field Testing

Determining transmissivity or the storage coefficient by any means other than actual performance tests in the field is expensive, time consuming, and of questionable accuracy. Field-testing methods for determining these values have been developed and are thoroughly documented. These methods involve the application of a regulated stress (pumping) to the formation and the measurement of the effects (changes in water level) produced. The data are then treated mathematically to determine the factors (transmissivity and storage coefficient) that control the responses measured.

To obtain the required data, it is essential that, in addition to a pumping well, one or more nearby wells tapping the same aquifer be available for use as observation points. No set number of wells is required, but the more there are, the less likelihood there is of making an erroneous analysis. It is common practice to install a number of small-diameter test wells in an area of investigation for this purpose. The location of all wells is, of course, accurately plotted on the area map so that the lateral distance and direction from the pumping well and the relative position with respect to other wells can be readily noted.

Water-Level Measurements

It is of value to set a benchmark and relate the elevations at all wells; then, by accurately measuring water levels with respect to surface elevations, groundwater gradients can be detected. For this purpose and for the collection of water-level data during an aquifer performance test, a definite reference point on the casing should be established. All measurements to water levels are made from that point. It is important to set up data sheets that adequately identify each well by number or other description. When a water-level measurement is made, the date, time, and distance to water should be recorded.

Tape method. Water levels are measured most simply using a hand-held tape with a weight attached to the end to hold it straight and taut. The tape should be metal, and graduated in feet and in tenths and hundredths of a foot, or in metric units. Such graduations facilitate calculations by eliminating conversion of fractions to decimal equivalents, which tend to become unwieldy. By chalking the lower portion of the tape and lowering it into the water until an even foot graduation coincides exactly with the reference point, it is easy to determine by simple subtraction the precise distance to water from the reference. The wetted chalk is easily identified by comparison with the dry area, and direct readings to one hundredth of a foot can be made. A little practice quickly gives an observer adequate experience.

Other methods. Other methods of collecting water-level data include an electric tape that has an insulated wire with an open-end weighted electrode on the end. When the electrode enters the water, it completes a circuit that actuates a light, buzzer, meter, or other signal device. The distance to water is then read directly from graduations on the wire line. However, the graduations are not usually fine enough to permit a very accurate reading without some supplementary device. Float-actuated recording devices provide a means of collecting data continuously, but the time drive is not fast enough for the early periods of a test program. Air-line devices have little value except where water-level fluctuations are very large.

COLLECTION OF TEST DATA

Test data must be collected and recorded carefully. Because water-level data are commonly plotted on a logarithmic time scale, the measurement increments should coincide with the plotting technique.

Collection Schedule

A collection schedule that can be easily followed and that provides adequate data is charted below.

1 reading at zero time	total elapsed time = 0 min
1 reading each 1 min for 10 min	total elapsed time = 10 min
1 reading each 2 min for 10 min	total elapsed time = 20 min

1 reading each 5 min for 20 min	total elapsed time = 40 min
1 reading each 10 min for 60 min	total elapsed time = 100 min
1 reading each 20 min for 80 min	total elapsed time = 180 min

All times are calculated from the precise instant that the pump is turned on or off, which is designated as zero. If the test extends beyond 24 h, subsequent measurements can be made at about 4-h intervals. It is evident that the timing of measurements at the onset of the test is critical; therefore, at each well there should be at least one observer equipped with measuring devices and a stopwatch synchronized with all others being used. After 180 min, it is not necessary to make the measurements at a designated instant, but an accurate record for the exact time of each measurement should be maintained. The necessity for and value of such data become apparent as the analysis procedures are pursued.

ANALYSIS PROCEDURES

The following discussion of some of the most popular procedures for analyzing aquifer test data is adapted from Brown (1953). Other references are also available (see the bibliography at the end of this manual).

Preliminary Considerations

All methods discussed here are designed to yield information on aquifer performance, not well performance. Each method will involve turning a pumped well on or off and observing what happens to the water level in nearby observation wells. All methods involve the use of the Theis nonequilibrium formula or modifications thereof. The name is derived from the man who first developed it and from the fact that it takes into account the time that has elapsed since pumping began or ceased.

Ideally, all wells that are used for the analysis should fully penetrate the aquifer. Some departures from this requirement can be tolerated, but the construction details of the wells used should be known or determined. Operations of any pumps in the area, not involved in the test, should be stabilized before an aquifer test is begun and maintained that way for the duration of the test. During the test, well pumping should be at a steady unvarying rate, carefully measured. Pumping rate and water-level data should be carefully computed and plotted.

Finally, each method will relate to those uses of the Theis formula that consider variations in drawdown with the passing of time or variations in drawdown with distance from the pumped well.

Hypothetical Test Setup

A demonstration of some of the test and analytical methods is facilitated by use of a hypothetical test setup. It will be convenient to consider a physical situation as shown in Figure 4-1. This illustration depicts a sand aquifer confined above and below by relatively impermeable clay. It is proposed to pump one well at 500 gpm and observe water-level changes in two nearby wells, wells 1 and 2. The wells fully penetrate the aquifer and, insofar as can be determined from the sectional view, the aquifer extends laterally to infinity. Infinity, in this case, means the aquifer is extensive relative to the effects of pumping. There is no evidence of nearby wells whose pumping might affect the test result. The observation wells could have been located anywhere in the general vicinity of the pumped well, but, for convenience, they were placed in a straight line.

58 GROUNDWATER

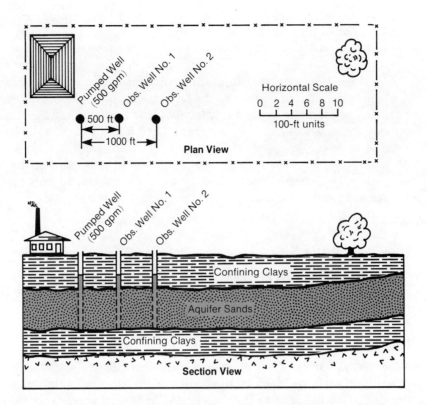

In this hypothetical situation, one well will be pumped at the rate of 500 gpm, and water-level changes will be noted in observation wells 1 and 2.

Figure 4-1 Hypothetical test situation—infinite aquifer.

A family of type curves has been developed to facilitate aquifer evaluation under a variety of conditions. The basic formulas are

$$T = 114.6QW(u)/s \qquad \text{(Eq 4-1)}$$

Where:

T = the transmissivity of the aquifer, in gallons per day per foot
Q = the discharge rate of the well, in gallons per minute
u, for any given formation, is proportional to the ratio of r^2/T
$W(u)$, the "well function of u," is determined from calculated tables from each value of u
s = the drawdown at any point under study in the vicinity of the discharging well, in feet.

$$u = 1.87r^2S/Tt \qquad \text{(Eq 4-2)}$$

Where:

r = the distance from the discharging well to the point where the drawdown is being observed, in feet
S = the aquifer storage coefficient
T = the transmissivity of the aquifer
t = the elapsed time since discharge began, in days.

Confined aquifers. A confined, or artesian, aquifer is confined above and below by relatively impermeable materials. The aquifer is homogeneous and isotropic-uniform in structure and with the same physical and hydraulic properties in all directions. In practical terms, the thickness and actual extent of the aquifer should be known so as to permit the best possible interpretation of the test data.

Leaky aquifers. The modified nonequilibrium formula for leaky artesian conditions is based on several assumptions, in addition to those that apply to nonleaky conditions as described above. These assumptions are (1) the aquifer is confined between an impermeable bed and a bed through which leakage can occur; (2) leakage is vertical into the aquifer and proportional to the drawdown; (3) no water is stored in the confining bed; and (4) the hydraulic head in the deposits supplying leakage remains constant.

Unconfined aquifers. An unconfined, or water-table, aquifer does not have water confined under pressure beneath impermeable rocks. Water is derived from storage by gravity drainage of the interstices above the cone of depression, by compaction of the aquifer, and by expansion of water in the aquifer.

Properties of an unconfined aquifer can be determined by the Theis method under some limiting conditions. One of the basic assumptions of the Theis solution is that water is released from storage instantaneously with a decline in head. In a water-table aquifer, this is not always true, because water is derived partly from gravity drainage, and the effects of gravity drainage are not considered in the Theis formula. However, with long pumping periods, the effects of gravity drainage become negligible so that the Theis solution can be used.

Drawdown Method

As its name implies, the drawdown method involves pumping one well, in a hypothetical test setup, and observing what happens to the water levels in two or more nearby wells. It might be helpful to illustrate what may be expected to happen in one of the observation wells during the drawdown test and to demonstrate what is

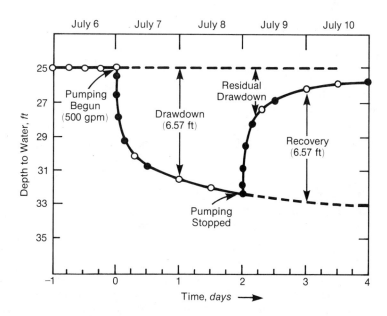

Drawdown data are plotted on the left curve, recovery data on the right. These data are for observation well 1, located 500 ft from the pumped well. Points indicated by O are used in later analysis plots. Arrows indicate directions of increasing scale values.

Figure 4-2 Hydrograph for observation well 1.

Table 4-1 Drawdown Test Data for Observation Wells

Date (July 1989)	Hour	Elapsed Time min	t (days)	$\dfrac{t}{r^2}$	Drawdown* s (ft)	Depth to Water ft
	Well 1 (r = 500 ft)					
5	2400					25.00
6	0600					25.00
	1200					25.00
	1800					25.00
	2400†	0	0	0	0	25.00
7	0004	4	0.00278	1.1×10^{-8}	0.44	25.44
	0015	15	0.0104	4.2×10^{-8}	1.50	26.50
	0055	55	0.038	1.5×10^{-7}	2.83	27.83
	0305	185	0.13	5.2×10^{-7}	4.22	29.22
	0600	360	0.25	1.0×10^{-6}	4.96	29.96
	1200	720	0.50	2.0×10^{-6}	5.75	30.75
	2400	1440	1.0	4.0×10^{-6}	6.57	31.57
8	1200	2160	1.5	6.0×10^{-6}	7.04	32.04
	2400	2880	2.0	8.0×10^{-6}	7.32	32.32
	Well 2 (r = 1000 ft)					
5	2400					25.10
6	0600					25.10
	1200					25.10
	1800					25.10
	2400†	0	0	0	0	25.10
7	0030	30	0.0208	2.1×10^{-8}	0.89	25.99
	0155	115	0.080	8.0×10^{-8}	2.16	27.26
	0640	400	0.278	2.8×10^{-7}	3.53	28.63
	2400	1440	1.0	1.0×10^{-6}	4.94	30.04
8	2400	2880	2.0	2.0×10^{-6}	5.75	30.85

*Values in this column are derived from the depth-to-water measurements made in the observation well and given in the next column.
†Pumped well begins discharging at 500 gpm.

meant by the term drawdown. Figure 4-2 is a hydrograph—a plot of water level versus time—for observation well 1 (Table 4-1). Only the left half of Figure 4-2 should be considered at this point. Water-level measurements were taken for a day prior to the start of the test in order to determine whether any preexisting upward or downward trend would have to be considered during the test. Again, for convenience, it was assumed that there was no upward or downward trend of water levels in the area, and the measurements are shown plotted along a horizontal line. Referring to the portion of the hydrograph after pumping started, one sees that the drawdown represents the difference between the water level observed in the well and the level at which the water would have stood had no pumping occurred. In the drawdown method, similar data will be collected for the observation wells and analyzed as the test proceeds. In the analysis, two kinds of solution are available.

Type-curve solution. As already stated, the Theis nonequilibrium formula is the basis for all the analytical methods to be considered here. Analysis permits the determination of aquifer transmissivity and storage coefficient by comparing a logarithmic curve of time versus drawdown against one of a series of type curves

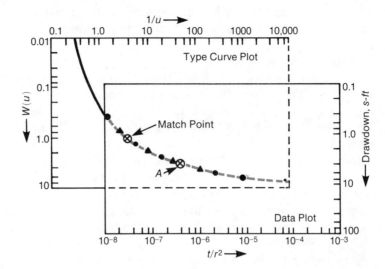

Figure 4-3 Drawdown test data superposed on Theis-type curve.

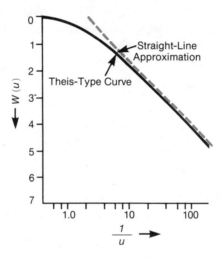

Figure 4-4 Straight-line approximation of drawdown data analysis.

developed from interpretation of the Theis formula. In comparing the field-data curve and the type curve, the type curve is superposed over the field-data plot, keeping the respective graphical axes parallel. The curves are adjusted horizontally and vertically so as to obtain the best match of the two curves. An arbitrary match point is selected on the two graphs, permitting selection of convenient field-curve and type-curve coordinates for substitution in the appropriate equation (Figure 4-3).

A different form of the type-curve solution is the distance-drawdown method. In this analysis, drawdown in three or more observation wells at different distances from the pumped well is compared with another interpretation of the type curve.

Detailed examples of analysis and variations of the type-curve form of solution are not given here. The interested reader should review the references cited (see the bibliography at the end of this manual). Scientists in the field of groundwater hydrology may develop individual preferences for specific analytical methods, but the fundamental principles and theory are common to all. The particular method favored will often be governed largely by the physical setup for collecting data. Development of computer programs have provided rapid advances to assist in the analysis of well-test data.

Straight-line solution. A second form of solution available for analyzing aquifer test data is really an abbreviated and approximate version of the type-curve solution. Plotting well-test data on semilogarithmic paper and using variations of the basic formula permits computation of the aquifer transmissivity and storage coefficient. The drawdown data tend to follow a straight line when plotted on semilog paper (Figure 4-4).

Identification of aquifer boundaries. What happens to the drawdown test data plot if an aquifer is not infinite, but rather has identifiable boundaries? This scenario is discussed in the following paragraphs.

Impermeable-barrier effect. To demonstrate the effect of an impermeable barrier around an aquifer requires another look at the plan and section views of a hypothetical situation (Figure 4-5). Everything is as it was illustrated in Figure 4-1, except that, in the right-hand direction, the aquifer does not extend out to infinity, but is cut off by an impermeable barrier in the form of the rising side of a buried valley. This situation is quite common in the northern, once-glaciated parts of the United States. Indeed, it often happens that an aquifer is cut off in two parallel directions by such buried-valley walls. It is enough, for the purpose of this discussion, to view the effects of a single barrier. For convenience, the effects will be analyzed using the straight-line solution of drawdown analysis.

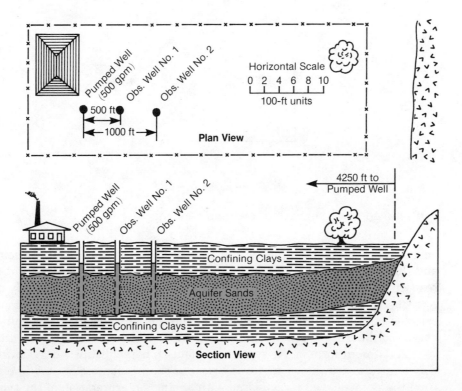

The situation is the same as in Figure 4-1, except that the aquifer is cut off by an impermeable barrier on the right side.

Figure 4-5 Hypothetical test situation—aquifer bounded by impermeable barrier.

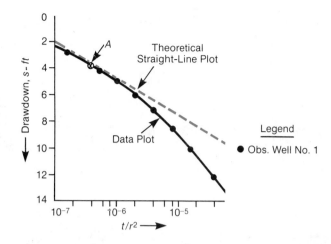

The drawdown data depart from the theoretical straight-line plot because the impermeable barrier limits the extent of the aquifer and increases the drawdown rate.

Figure 4-6 Effect of impermeable barrier shown on straight-line drawdown plot.

The early data arrange themselves in the expected manner, with a curved portion leading into a straight line. However, instead of staying on a straight line, the plotted data now curve off below it and eventually define a new straight line having twice the slope of the original (Figure 4-6). In other words, drawdown in the observation wells occurs at a faster rate than if the aquifer were of infinite extent. The effect is the same as that of a well identical to the pumped well, located across the boundary at the same distance and pumping at the same rate (image well theory).

It is possible to take these data and determine, not only the presence and kind of aquifer boundary, but an aquifer average position with respect to the pumped well. Detailed explanation of these procedures is given in numerous references (see bibliography at the end of this manual), and is beyond the scope of this discussion.

Recharge effect of local stream. One other aquifer boundary effect that is of interest is that of a recharging stream. Figure 4-7 shows the same hypothetical situation, except that the aquifer is cut off on the right-hand side by a recharging stream—a situation that is often found in the field.

As before, it will be convenient to see how the recharge effect of this stream is exhibited on the straight-line form of a drawdown plot (Figure 4-8). Once again the plot starts out as expected, with a curved portion leading into a straight line near point A. Instead of continuing on the straight line, as the data theoretically should for an infinite aquifer, the plotted data curve away above it and eventually define a horizontal line. Thus, the rate of drawdown slackens, because of the water contributed to the aquifer by the stream, and gradually approaches a fixed value. This effect is the same as if a well, identical to the pumped well, was recharging the aquifer at an equal distance from and on the opposite side of the recharge boundary.

From these data, it would be possible to interpret not only the presence and nature of the aquifer boundary, but also its location with respect to the pumped well. A complete discussion of these methods is given in many texts on the subject.

Recovery Method

A second important method of analyzing aquifer test data is the recovery method, which involves shutting off a pumped well and observing the recovery of water levels

64 GROUNDWATER

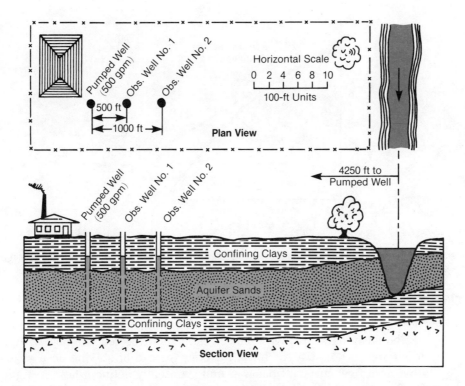

The situation is again the same as in Figure 4-1, except that the aquifer is bounded on the right by a recharging stream.

Figure 4-7 Hypothetical test situation—aquifer bounded by recharging stream.

in nearby observation wells. In considering the types of solutions available, it is desirable to reexamine Figure 4-2 to see how recovery is measured. Recovery is the difference between the observed water level in the well, some time after pumping has stopped, and the level at which the water would have been had pumping continued. It can be noticed on the hydrograph that one day after pumping stopped, there was a recovery of 6.57 ft, which exactly equaled the drawdown observed one day after pumping began.

Type-curve solution. When using the type-curve solution, the same curve is used as in the drawdown method, except that it has been inverted to indicate a rising trend comparable to that anticipated for the water levels in the observation wells. A plot of recovery measurements for the observation wells is an upside-down version of the drawdown plot. Comparing the recovery curve with the inverted type curve permits the determination of values for the transmissivity and storage coefficient, which should be similar to those obtained using the drawdown method of analysis.

Straight-line solution. As in the drawdown method, both the type curve and the data curve are plotted on semilog paper when using the straight-line solution. The type curve is inverted to show a rising trend in the recovery period. With these modifications, the curves become straight lines. The same abbreviated equations permit computation of the transmissivity and storage coefficient with results similar to those achieved in the drawdown method.

Of the two basic methods for analyzing aquifer test data discussed so far—the drawdown method and the recovery method—each permit two kinds of solutions—a type curve, or graphical solution, and a straight-line, or approximate solution. It has been shown that the drawdown solutions closely parallel the recovery solutions. In

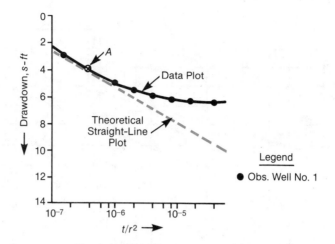

Instead of following the theoretical straight-line plot, the drawdown data curve shows an upward trend because the recharging stream replenishes the aquifer, reducing the drawdown rate.

Figure 4-8 Effect of recharging stream shown on straight-line drawdown plot.

practice, therefore, it is not necessary to use two kinds of type curves and two kinds of straight-line plots. If it is recognized that a recovery test is essentially the reverse of a drawdown test, one type curve and one straight-line plot will serve equally well for either kind of test data. In fact, both kinds of data can be recorded on the same plot to check their agreement.

Specific-Capacity Method

An abbreviated well-performance evaluation can be performed using a relatively short test to determine the specific capacity of the well. Specific capacity is defined as the yield of the well, usually expressed in gallons per minute per foot of drawdown (gpm/ft [m^2/day]). The total drawdown in a well can be divided into two components—drawdown in the aquifer and drawdown related to well loss. Drawdown in the aquifer depends on the aquifer's ability to transmit water, and generally does not change unless the aquifer is being depleted. Drawdown due to well loss is related to the ability of the well to transmit water, and may change with time, due to turbulent flow or head loss as the water passes through the screen or well bore. Some of the factors affecting well loss include changes in chemical or bacterial quality of the water, or changes in the mechanical condition of the well itself.

Monitoring specific capacity is a valuable tool for helping detect well-maintenance problems before they become critical. Tests for specific capacity should not be substituted for the more involved tests described above when it is necessary to make a more complete well and aquifer evaluation.

WELL-FIELD DESIGN

As Brown (1953) indicates, the basis is available for the proper design of wells and well fields through measurable field data. The individual water system requirements, area development, geology, hydrology, and climatology must all be considered. More specifically, an aquifer's hydraulic properties—transmissivity and storage coefficient—may be used in a number of ways. Determination of the most desirable spacing between wells in a field, the effects of new wells on existing wells, and the

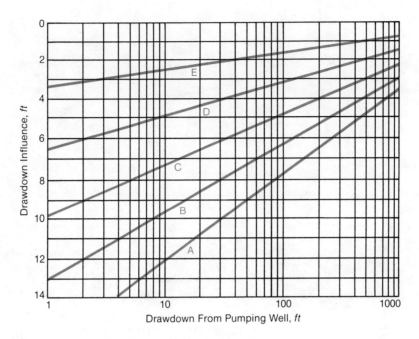

Aquifer has a *T* value of 60,000 gpd per foot and an *S* value of 0.0055. The *Q* values (rate of pumping, in gallons per minute) for curve A is 500 gpm; curve B, 400 gpm; curve C, 300 gpm; curve D, 200 gpm; and curve E, 1000 gpm.

Figure 4-9 Influence for various rates of pumping in an aquifer.

optimum pumping rates and schedules can be made once the thickness and extent of an aquifer, its transmissivity and storage coefficient, and the nature and location of boundaries are known. Furthermore, this knowledge is of considerable value when making an overall appraisal of the groundwater resources of an area and the potential for future water supply development. All of this information will govern the extent to which design calculations must be used. In all cases, these factors should be considered in a logical order.

Pumping Rates

Once numerical values have been assigned to transmissivity and the storage coefficient, it is possible to determine the drawdown effects of pumping any quantity of water in the vicinity of and at any reasonable distance from the pumping well. Common practice is to plot a graphic representation (Figure 4-9) of water levels against the logarithm of distance from the center of pumping for a given time period. To incorporate a logical safety factor and a degree of conservatism, a minimum continuous pumping period of 100 days is usually chosen for such purposes.

Well-Field Interference

In addition to determining the optimum pumping rate for individual wells, analysis can be of considerable value in determining interference between wells in a well field. Use of this procedure prior to locating individual wells or multiple wells in a well field will result in the most efficient placement pattern and pumping rates.

As an example, assume that the data given in Figure 4-5 are representative of the hydraulic conditions. The available land for a well field measures 600 ft on a side.

Local health department regulations require a well distance of 200 ft from property lines. The two most advantageous distributions of wells appear to be nine wells 100 ft apart or four wells 200 ft apart. From the results of drilling, testing, and calculations, the probable drawdown in the vicinity of a well can be determined for a given pumping rate. The difference between the total available and the calculated drawdown represents the allowable interference drawdown. For the sake of illustration, it shall be assumed that the total interference drawdowns that can be accommodated for various pumping rates are as shown in Table 4-2.

Table 4-2 Allowable Interference Drawdowns for Various Pumping Rates

Pumping Rate gpm	Probable Self Drawdown ft	Allowable Interference Drawdown ft
100	4.4	31.6
200	8.6	27.4
300	12.7	23.3
400	17.2	18.8
500	21.0	15.0

The problem of design, then, is one of balancing the cost of well and pump installation against the quantity of water produced to get the best returns. An installation with nine wells and pumping rates of 200 gpm each will serve as an example.

As indicated in Table 4-2, the total interference must not exceed 27.4 ft; in the nine-well pattern, this will present the total effects of eight other wells. A corner well will be the least affected by the pumping of its companion wells, and the center well will suffer the greatest effects. Therefore, consideration of the latter will give the quickest appraisal of the problem. For the center-well case, the combination of four wells at 100 ft distance and four wells at 141 ft distance will represent the total interference. Referring to the graph in Figure 4-5 and following the 200-gpm line, one can determine that wells at 100 ft will have an influence of 3.1 ft per well and wells at 141 ft will have an influence of 2.9 ft per well. The total interference expected would be 12.4 + 11.6 = 24.0, which is below the 27.4 ft maximum allowed over the 100-day pumping period.

A four-well configuration with wells rated at 500 gpm would require the addition of influences from two wells at 200 ft and one well at about 283 ft. The total influence in that case would equal 18.8 ft, which is above the allowable limit. It may be of interest to know that 2000 gpm could be produced from this field from a square of eight wells, each pumping 250 gpm. No center well would then be involved. The most practical solution, however, would probably be four wells pumping about 425 gpm each.

Any example can be handled in a similar manner if the T and S values are known and other variables can be reasonably approximated. For very large areas, it may be necessary to use models and computer calculations. It should be further noted that a number of field-performance tests for data collection to evaluate T and S for such use increases the reliability of calculated withdrawal effects.

WELL LOSSES

Drawdown values obtained for a single pumping well through straight application of the Theis formula represent only the head losses suffered by water movement through the formation under laminar flow conditions. The actual pumping level of a particular well cannot be calculated without additional considerations for high velocities and turbulence losses as a result of its own pumping. At and near the well face, fluid velocities usually become so large that turbulent flow conditions exist. The magnitude of turbulence losses varies with each well because of differences in formation characteristics, screen slot sizes required, degree of well development, well diameter, and quantity of water being pumped. There are so many unknown quantities involved in the calculation of these individual factors that they are usually lumped together under the heading of "well losses."

Calculation

A method of approximating the well losses for a particular well has been presented by Rorabaugh (1953) as follows:

$$s_w = BQ + CQ^2 \qquad \text{(Eq 4-3)}$$

Where:

s_w = observed drawdown in the pumped well
B = the coefficient of formation losses
C = the coefficient of well losses
Q = the pumping rate.

The values of B and C may be calculated if proper test data are available. To collect such data, it is necessary to pump the finished well at three to five progressive rates for equal periods of time and observe the development of drawdown in the pumping well at each pumping rate. When a full-scale aquifer performance test is not conducted, however, a step drawdown test offers a method of breaking down the observed losses in the pumping well. Additionally, this test makes it possible to quickly compare the magnitude of well losses to determine when a well is in need of cleaning or other repair work. Increasing well loss with increasing pumping rates indicates unsatisfactory development of a new well, or deteriorating aquifer or well conditions in an old well.

A controllable factor that plays a significant part in the magnitude of well loss for sand and gravel wells is open screen area. Research indicates that fluid velocities through the screen openings between 0.1 and 0.2 ft/s (0.03 and 0.06 m/s) permit satisfactory extraction performance. No penalty is suffered for velocities lower than 0.1 ft/s (0.03 m/s), but velocities greater than the 0.2-ft/s (0.06-m/s) limit may result in higher pumping and maintenance costs. In this case, velocity is a function of quantity and area and is easily approximated in the design stages by following the method outlined in chapter 3.

Because the quantity of water to be pumped from a well Q is more correctly established by consideration of formation loss and well interferences, the open area of screen is the basic consideration to be made here. Screen slot size should be determined by accurate sampling and proper sieve analysis. Thus, the screen diameter and length remain as the two variables in design.

In choosing a supply-well diameter, one should consider the pumping equipment that will be installed in the final well. It is usually not wise to install an 8-in.

(200 mm) turbine pump in an 8-in. (200-mm) diameter well or a 10-in. (250-mm) pump in a 10-in. (250-mm) well. Most pump installers prefer at least 1-in. (25-mm) clearance between the pump bowls and well casing. This permits adequate space for pump installation and removal, efficient pump operation, and good hydraulic efficiency of the well.

The question of screen-length selection should incorporate more than a casual recollection of the aquifer thickness. The definition of T incorporates flow through the total thickness of water-bearing material. If less than the total thickness is used, it can be expected that the value of T should be decreased. The Theis equation indicates that as T decreases, the formation drawdown will increase. Fortunately, this is not a directly proportional increase. But, if the screened portion of the formation is significantly less than one half of the formation thickness, the additional drawdown suffered may be significant. Therefore, as much of the aquifer as practical should be screened to eliminate losses of yield.

RADIAL-WELL YIELD

A description of a radial well is given in chapter 3. Yields of these types of wells are dependent on a permeable aquifer, a high water table, and an adequate, nearby source of water of acceptable quality, as well as suitable design characteristics to allow the desired volume of water to enter the gallery and, at the same time, retain fine-grained aquifer material outside the installation. Entrance velocities through the screen slot openings should average about 0.1 ft/s (0.03 m/s) or less. Capacities of collectors can vary considerably, depending on aquifer characteristics and well design. Yields may range upwards to several hundred gallons per minute (700 to 800 litres per minute) per 1000 ft (300 m) of gallery length.

References

BROWN, R.H. Selected Procedures for Analyzing Aquifer Test Data. *Jour. AWWA*, 45:8:844 (Aug. 1953).

RORABAUGH, M.I. Graphical and Theoretical Analysis of Step Drawdown Test of Artesian Well. *Proc. ASCE*, 362:79 (Dec. 1953).

AWWA MANUAL M21

Chapter 5

Use of Computer Models in Groundwater Investigations

Man places a variety of stresses, such as pumping and irrigation, on groundwater systems. These stresses produce a wide variety of complex responses in aquifer systems. Awareness of these stresses has fostered the development of computer models to aid in the analysis of groundwater conditions and evaluation of system responses. The use of computer models to analyze groundwater problems has increased significantly in recent years. The computational speed made possible by modern computers makes solving groundwater problems easier and more efficient. Hydrologists interested in evaluating effects of stresses on an aquifer system will find computer models very useful, as will managers who make decisions related to efficient use of groundwater resources (Figure 5-1).

However, for both the individual water supplier and the small-community water supplier, commissioning a model study to help determine where to locate a well and its pumping schedule may not be economical. The individual or local supplier may be unable or unwilling to spend thousands of dollars to model their problem. He or she may, however, have access to a model at a relatively modest cost. Such models are typically maintained by universities or extension services. Nonetheless, the use of models at the individual and local levels for small water problems is uncommon.

At present, computer models are most widely used at the area and regional levels for management decisions. The government agencies at these levels and the groundwater issues with which they deal are of a scale that makes computer model-

Figure 5-1 Computers are commonly used today to model groundwater conditions.

ing economically viable. Most groundwater modeling performed at these levels pertains to understanding the groundwater flow system and to hydraulic considerations affecting feasibility, investment, and operations decisions related to particular site-specific problems.*

MODELS OF THE GROUNDWATER SYSTEM

Computer models for groundwater systems can be divided into the following groups:

- aquifer response models—simulate the behavior of the groundwater system and its response to stress, enable evaluation of impacts of the stresses;
- resource management models—integrate the behavior of an aquifer response model with explicit water management decisions;
- parameter identification models—sometimes termed inverse models, help determine model-needed input parameters on the basis of available historical information; and
- data manipulation models—process and manage data.

At the simplest level, the models yield analytical solutions, if the aquifers are assumed to be isotropic and homogeneous with simple flow and regular geometrical boundaries. In complex situations, numerical techniques are used to solve complex mathematical equations, which describe groundwater flow in anisotropic and heterogeneous aquifers with complex boundaries.

---

*The material that follows in this chapter has been adapted from Bachmat et al. (1978). A more complete explanation of the use of groundwater models for water resource management is found in that report.

Aquifer Response Models

Aquifer response models are the most commonly used groundwater models today. They can be subdivided into four major categories: flow, mass transport, heat transport, and deformation (subsidence). The purpose of any of these models is to evaluate spatial and temporal changes in the movement of water, contaminants, heat, and land due to stresses.

A common feature of any aquifer response model is its deterministic nature (a single value) rather than probabilistic nature (a range of value of varying probability). In other words, aquifer response models are used to solve for one value rather than a range of values. In most cases, the model results are unconditional, implying that the model does not contain any restrictions. However, some aquifer response models have provisions for including operating rules, as well as constraints on water levels, water quantity, and/or water quality. Such models are useful in evaluating effects of alternative policies of groundwater-related management of water resources.

Some aquifer response models, especially flow models, contain nonhydrologic information, such as crop production. These models can be used to assess the hydrologic effects of nonhydrologic activities and vice versa. Such models are also used in conjunction with optimization techniques in management models.

Flow model. A flow model is the most commonly used, as well as the best developed, of the aquifer response models. Almost any modeling activity related to groundwater starts with a flow model. Since water quality and quantity are intimately related, flow information is essential to any water quality or deformation model. Therefore, many other types of models incorporate a flow model within them. Flow models have also been developed for the analysis of the unsaturated zone.

Purpose. A flow model serves a variety of purposes. The model uses information regarding aquifer parameters, boundary conditions, and stresses as input data when evaluating the quantitative aspects of groundwater behavior. Most models are used to evaluate the distribution of heads and rates of flow in aquifers under differing patterns of recharge and discharge. Flow models may be used to study the interaction among aquifers and streams; location of boundaries between native and injected water bodies; travel of water particles during well injections; movement of the interface between fresh and salt water; and spring flow, drawdowns, and evapotranspiration. A flow model may be used in preliminary evaluations of contamination problems, if the contaminants are assumed to be conservative and move with water particles. In general, most flow models are applied to field problems, and many are sufficiently documented and flexible enough to be used by individuals other than those who developed them.

Most flow models are of the distributed type; that is, they have spatial components. Most common are two-dimensional models, which consider the flow in an aquifer as essentially horizontal. The most complex flow models are multilayered, three-dimensional flow models. These models handle flow in various kinds of aquifers: confined, unconfined, leaky, and nonleaky. Some of them even contain a submodel of vertical flow through the unsaturated zone. A three-dimensional model can also be used to evaluate head distributions with depths and the configuration of boundaries.

Limitations. Some of the limitations of present flow models are worth noting. Although there is a relatively large number and variety of models available today that handle flow in aquifers and in the unsaturated zone, relatively few can integrate both groundwater and surface water conditions. Most flow models do not handle moving boundaries or flow in aquifers dominated by fractures, joints, or solution caverns. The major technical difficulties with flow models are obtaining solutions, in the case of drastically varying flow conditions, and trying to find an appropriate compromise between the desired level of model accuracy and the amount of data and computer capacity available.

Mass transport models. A mass transport model deals primarily with groundwater quality. It is used to evaluate the concentration and movement of various aquifer pollutants, including radionuclides, leachate from landfills and irrigated areas, and salt water intruding in coastal areas. The model incorporates mathematical approximations of the transport of one or more pollutants in the groundwater, by means of fluid flow. A transport model that describes the movement of pollutants without alteration of their identity and quantity is termed a conservative model; a nonconservative model does not perform this function.

In principle, a mass transport model contains a flow model, which computes the flow velocities. The transport model considers transport of the contaminant in the flow field; this allows for additional spreading or mixing (dispersion) and transformations by reactions, and/or adsorptions on or from rock minerals. Under certain circumstances, such as low contaminant concentrations, flow and transport models can operate independently. In other cases, however, their mutual effects cannot be separated. Thus, relatively high contaminant concentrations in wastewater/salt water or contaminant density can affect the groundwater flow pattern, which affects the movement and spread of the contaminant. The high contaminant concentration requires a more intimate interactive coupling between the flow and transport models. In conclusion, it is fair to say that today's mass transport models are adequate for obtaining preliminary estimates of contaminant movement.

Heat transport models. A heat transport model couples the flow of heat with water or steam for problems in which thermal effects are important. A heat transport model is conceptually similar to a mass transport model in that a fluid flow model is coupled with a heat flow model, which incorporates various mechanisms of heat transfer.

In practice, these models have been applied to problems associated with hot springs, geothermal reservoirs, and underground heat storage. The temperature of groundwater is usually of lesser importance to the water resource manager than its chemical or bacteriological quality, making heat transport models less useful than other types of models. However, heat transport models have recently grown in importance as underground storage of heat, thermal pollution, and exploitation of geothermal energy have become more common.

Deformation models. Deformation, or subsidence, models describe the phenomenon of land subsidence caused by excessive withdrawal of fluids from an aquifer. These models are needed to evaluate deformation-related impacts due to human activities. Usually the land surface deformation is computed on the basis of heat change (pressure change) and the amount of water released from thick clay beds obtained from a flow model under various conditions.

Resource Management Models

Resource management models have been developed to indicate courses of action that will be consistent with stated management objectives and constraints. The objectives may be, for example, to maximize economic benefits, to minimize costs, or to ensure adequate water supply at all times for all users. A resource management model deals primarily with engineering decisions and a single economic or physical objective. They are conceptually more complex than regular models, because they provide for additional aspects of decision making. Management models may use the techniques of both simulation (characterizing an actual situation) and optimization (solving for defined goals). In contrast to purely physically based mathematical models, management models incorporate economical, technological, political, and institutional aspects of the problem being analyzed.

Management models incorporate many different types of models and data, as well as additional computations. Usually a management model contains four elements: a model for finding the most appropriate decisions (for example, location of well fields, pumping rates); a model for evaluating the outcome of a decision (for example, water-level declines, salinity); a set of rules and constraints on actions and/or outcomes (for example, maximum pumping rates, critical drawdowns, degree of salinity, water rights, well regulations); and a so-called objective function, which optimizes the decision (for example, least cost, maximum benefit, and yield).

Quantity and quality management. Models exist for managing the quantity and quality of groundwater or both groundwater and surface water. The combined management model for groundwater and surface water quantity is distinctive in that it considers a variety of multicomponent systems, such as the amount of water resources, supply and use, production, and management tasks at the regional level. Most combined models actually deal with quantity management through two types of models: lumped models and distributed models.

Some lumped models seek to optimize the price of water, taking into consideration the relationship between demand, supply, and price. Other lumped models seek to optimize development and operation, while either maximizing expected benefits or minimizing costs.

All groundwater management models are of the distributed type, and most models deal with water quantity management in a single aquifer containing a well field. Usually the objective is to distribute the pumpage so as to satisfy a given demand at minimum cost or to maximize benefits from the use of the pumped water. Other models address decisions for agricultural production, fiscal policies to encourage efficient use of groundwater, and cost–benefit evaluation of collecting additional data.

Some distributed quantity-management models are distinctive in that they treat stream–aquifer interactions and address coordinated multilevel management. The latter is a developed methodology that can serve the needs of decentralized water resource management. Through this methodology, several independent agencies that administer different parts of a regional water resources system can coordinate their activities so as to best use the limited water resources of the region. Another advantage of this methodology is the decomposition of large-scale complex water systems into smaller ones, while seeking a regionally optimal scheme of operations and/or development.

Quality management of a groundwater–surface water system can be handled by some resource management models. Some lumped models determine an optimal pattern of pumpage, wastewater treatment, and resource allocation so as to meet a desired quantity and quality of the supplied water at a given reliability level. Another type of model, which is the distributed type, aims at operating available local surface and groundwater resources in conjunction with water importation so as to satisfy agricultural water demand, in terms of quality and salinity, at a minimum cost.

Parameter Identification Models

Estimating the hydraulic parameters to be used in a groundwater model is probably the major stumbling block to efficient use of already available models. Although engineering techniques have long existed for calculating parameters through aquifer tests, identification of regional groundwater system parameters must be augmented by regular observations of wells throughout the regional system. As a result, parameter identification models are being developed through the statistical analysis of long-term historical data of groundwater systems.

The purpose of a parameter identification model is to solve a problem that is the inverse of response with excitation, that is, given historical information as model input, to find the parameter values of the model that ensure the model output is as close to the observed output as is possible through statistical analyses. Most existing parameter identification models estimate parameters related to groundwater quantity only, primarily for distributed horizontal flow models.

Indirect and direct approach. Two methods of calibration for parameter identification models are commonly used—the indirect and direct approach. Using the indirect approach, a succession of model outputs is used to successively improve parameter values. For the direct approach, parameter values are computed from the given input and output data without using a succession of model runs. The direct approach is attractive because it usually saves computer time and its results do not depend on the initial guess of the parameters. Both approaches have been widely applied, and most models find the best values of parameters by optimization techniques. A few models, however, still employ trial-and-error methods.

Limitations. Despite the progress made in developing parameter identification models, some basic conceptual and methodological limitations remain. These include the weak definition of "best" parameter values and the poor selection of such values. Other issues include methods of updating parameter values as new information becomes available and addressing the uniqueness and dynamic character of parameter values. To conclude, although some parameter identification models have been made operational, they have not, in general, been widely applied to practical problems.

Data Manipulation Models

The difficulties involved in estimating parameters are closely linked with the more general issue of data collection for groundwater models. While models can be run with any amount of available data, the actual amount and quality of these data directly affect the reliability of the model's results. The task of collecting appropriate amounts of accurate data to ensure model reliability thus implies the need for a fourth class of computer model—a data manipulation model. This type of model can be used in various ways, including specifying data-collection procedures, designing data-collection networks, identifying critical data, and storing and processing data for use in models.

Each of the models just discussed uses data of various kinds, such as water levels, contaminant concentrations, pumpage, and recharge. These data, most of which are collected in the field, are commonly placed in primary data files. They are then processed in various ways, through sorting, interpolation, aggregation, and statistical analysis, before being placed in secondary data files for use in the modeling process. Many of these tasks are performed by data manipulation models. In addition, such models, with output that may be in the form of tables, contour maps, and plots, are used for producing status reports on water levels and aquifer quality.

Reference

BACHMAT, YEHUDA, ET AL. Utilization of Numerical Groundwater Models for Water Resource Management. USEPA Rept. EPA—600/8-78-012. Robert S. Kerr Envir. Res. Lab., Ada, Okla. (1978).

AWWA MANUAL M21

Chapter 6

Well Pumps and Pumping

Pumps are used to produce flow by transforming mechanical energy to hydraulic energy. Pump designs and applications are numerous, and energy specifications and ratings for pumps range from less than one to thousands of horsepower per pump. In order to understand pumps and how they work, it is helpful to understand some basic terminology. The following terms and definitions are supplied to help the reader better understand pumps and pumping.

Capacity. The rate of flow delivered by a pump, expressed in units such as gallons per minute, cubic feet per second, or barrels per hour. To calculate the power needed or the size of prime mover required to produce a desired capacity, one must determine the rate of flow and the total dynamic head.

Dynamic head. The resistance to flow produced by a system. Dynamic head is equal to the sum of static head, velocity head, and friction head.

Static head. The static suction head plus the static discharge head (Figure 6-1). To calculate static head, consider all measurements involved in pumping as being in a vertical direction and use the maximum drawdown as a reference. Measurements above this level are considered positive; those below are negative. The same measuring procedure can be used for both submersible and surface-mounted pumps.

Static suction head. Vertical measurement, in feet, of the distance from the water level in a well to the pump centerline.

Static discharge head. The distance measured vertically from the pump centerline to the water level in storage.

Velocity head. The vertical distance, or height, through which a buoy must fall freely, under the force of gravity, to acquire the velocity that it possesses. In most cases, the velocity head is small and can be ignored. Table 6-1 is convenient for determining the velocity head.

Friction head. The loss of energy resulting from motion of fluid along the inner surfaces of pipe and through fittings. With no change in elevation considered, friction head is the amount of head necessary to push fluid through pipe and fittings

WELL PUMPS AND PUMPING 77

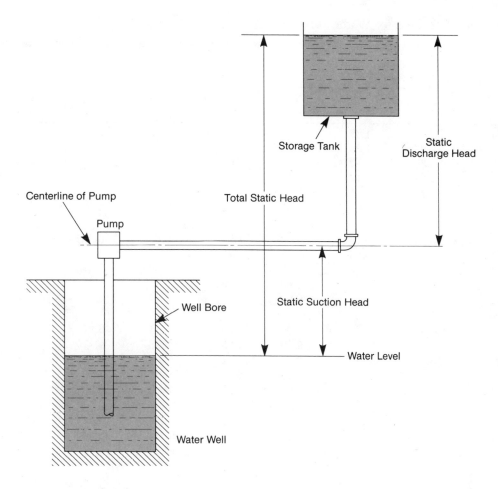

Figure 6-1 Schematic illustrating total static head.

Table 6-1 Velocity-Head Data

Velocity (V) fps	Velocity Head (h_v)* ft	Velocity (V) fps	Velocity Head (h_v) ft
1	0.02	11	1.87
2	0.06	12	2.24
3	0.14	13	2.62
4	0.25	14	3.04
5	0.36	15	3.49
6	0.56	16	3.97
7	0.76	17	4.44
8	1.00	18	5.03
9	1.25	19	5.61
10	1.55	20	6.21

*$h_v = V^2/2g$; g = acceleration due to gravity.

Table 6-2 Friction Loss for Water in Feet per 100 Feet (Schedule 40 Steel Pipe)

Flow gpm	2 in. V ft/s	2 in. h_f Frict.	2 ½ in. V ft/s	2 ½ in. h_f Frict.	3 in. V ft/s	3 in. h_f Frict.	4 in. V ft/s	4 in. h_f Frict.
25	2.39	1.29						
30	2.87	1.82						
35	3.35	2.42	2.35	1.00				
40	3.82	3.10	2.68	1.28				
45	4.30	3.85	3.02	1.60				
50	4.78	4.67	3.35	1.94	2.17	0.662		
60	5.74	6.59	4.02	2.72	2.60	0.924		
70	6.69	8.86	4.69	3.63	3.04	1.22		
80	7.65	11.4	5.36	4.66	3.47	1.57		
90	8.60	14.2	6.03	5.82	3.91	1.96		
100	9.56	17.4	6.70	7.11	4.34	2.39	2.52	0.624
120	11.5	24.7	8.04	10.0	5.21	3.37	3.02	0.877
140	13.4	33.2	9.38	13.5	6.08	4.51	3.53	1.17
160	15.3	43.0	10.7	17.4	6.94	5.81	4.03	1.49
180			12.1	21.9	7.81	7.28	4.54	1.86
200			13.4	26.7	8.68	8.90	5.04	2.27
220			14.7	32.2	9.55	10.7	5.54	2.72
240			16.1	38.1	10.4	12.6	6.05	3.21
260					11.3	14.7	6.55	3.74
280					12.2	16.9	7.06	4.30
300					13.0	19.2	7.56	4.89
350					15.2	26.1	8.82	6.55
400							10.10	8.47
450							11.4	10.65
500							12.6	13.0
550							13.9	15.7
600							15.1	18.6

Note: The table shows average values of pipe friction for new pipe. For commercial installations it is recommended that 15 percent be added to these values because no allowance for aging of pipe is included.

Table continues on next page.

at the required velocity. Table 6-2 can be used to determine friction head when various pipe sizes and different flow rates are used.

Friction head loss through fittings also must be considered and can be estimated by referring to Table 6-3 (pp. 82–83). In Table 6-3, head losses for fittings are expressed in equivalent feet of pipe. For example, the loss through a regular 4-in. 90° elbow is equivalent to the loss through 13 ft of 4-in. pipe at the measured flow rate.

To understand head loss, it is important to understand the relation between a pressure expressed in pounds per square inch (psi) and a pressure expressed in feet of head. This is expressed as

$$\text{head, in feet} = \text{psi} \times \frac{144}{w} \qquad \text{(Eq 6-1)}$$

Table 6-2 Friction Loss for Water in Feet per 100 Feet (Schedule 40 Steel Pipe) (continued)

Flow gpm	5 in. V ft/s	5 in. h_f Frict.	6 in. V ft/s	6 in. h_f Frict.	8 in. V ft/s	8 in. h_f Frict.
160	2.57	0.487				
180	2.89	0.606				
200	3.21	0.736				
220	3.53	0.879	2.44	0.357		
240	3.85	1.035	2.66	0.419		
260	4.17	1.20	2.89	0.487		
300	4.81	1.58	3.33	0.637		
350	5.61	2.11	3.89	0.851		
400	6.41	2.72	4.44	1.09	2.57	0.279
450	7.22	3.41	5.00	1.36	2.89	0.348
500	8.02	4.16	5.55	1.66	3.21	0.424
600	9.62	5.88	6.66	2.34	3.85	0.597
700	11.2	7.93	7.77	3.13	4.49	0.797
800	12.8	10.22	8.88	4.03	5.13	1.02
900	14.4	12.9	9.99	5.05	5.77	1.27
1000	16.0	15.8	11.1	6.17	6.41	1.56
1100			12.2	7.41	7.05	1.87
1200			13.3	8.76	7.70	2.20
1300			14.4	10.2	8.34	2.56
1400			15.5	11.8	8.98	2.95
1500					9.62	3.37
1600					10.3	3.82
1700					10.9	4.29
1800					11.5	4.79
1900					12.2	5.31
2000					12.8	5.86
2100					13.5	6.43
2200					14.1	7.02

Note: The table shows average values of pipe friction for new pipe. For commercial installations it is recommended that 15 percent be added to these values because no allowance for aging of pipe is included.

Table continues on next page.

Where:

w = specific weight, in pounds per cubic foot.

The specific weight of water at temperatures less than 85°F usually is taken as 8.34 lb/gal or 62.4 lb/ft^3, so each foot of water causes a change in pressure of 0.433 psi. To change from feet of water to pounds per square inch, multiply by 0.433 or divide by 2.307. For example, the pressure in pounds per square inch at the bottom of a storage tank containing a 10-ft depth of water is calculated

pressure, in pounds per square inch = 10 × 0.433, or

= 10 ÷ 2.307

= 4.33 psi

Table 6-2 Friction Loss for Water in Feet per 100 Feet (Schedule 40 Steel Pipe) (continued)

Flow gpm	10 in. V ft/s	10 in. h_f Frict.	12 in. V ft/s	12 in. h_f Frict.	14 in. V ft/s	14 in. h_f Frict.
650	2.64	0.224				
700	2.85	0.256				
750	3.05	0.294				
800	3.25	0.328				
850	3.46	0.368				
900	3.66	0.410	2.58	0.173		
950	3.87	0.455	2.72	0.191		
1000	4.07	0.500	2.87	0.210	2.37	0.131
1100	4.48	0.600	3.15	0.251	2.61	0.157
1200	4.88	0.703	3.44	0.296	2.85	0.185
1300	5.29	0.818	3.73	0.344	3.08	0.215
1400	5.70	0.940	4.01	0.395	3.32	0.217
1500	6.10	1.07	4.30	0.450	3.56	0.281
1600	6.51	1.21	4.59	0.509	3.79	0.317
1700	6.92	1.36	4.87	0.572	4.03	0.355
1800	7.32	1.52	5.16	0.636	4.27	0.395
1900	7.73	1.68	5.45	0.704	4.50	0.438
2000	8.14	1.86	5.73	0.776	4.74	0.483
2500	10.2	2.86	7.17	1.187	5.93	0.738
3000	12.2	4.06	8.60	1.68	7.11	1.04
3500	14.2	5.46	10.0	2.25	8.30	1.40
4000	16.3	7.07	11.5	2.92	9.48	1.81
4500			12.9	3.65	10.7	2.27
5000			14.3	4.47	11.9	2.78
6000			17.2	6.39	14.2	3.95
7000					16.6	5.32
8000						

Note: The table shows average values of pipe friction for new pipe. For commercial installations it is recommended that 15 percent be added to these values because no allowance for aging of pipe is included.

Table continues on next page.

Net positive suction head (NPSH). The amount of pressure required to prevent vaporization of the water that can cause cavitation (the formation and collapse of water vapor bubbles in the flowing water) and damage to a pump. The required or minimum NPSH usually is stated by the pump manufacturer. The available NPSH is approximately equal to the distance from the eye of the pump impeller to the water level in the well while pumping. The available NPSH must be at least equal to the required NPSH to prevent cavitation. If necessary, the required NPSH can be satisfied by lowering the pump in the well.

PUMP CLASSIFICATIONS

There are several types of pumps in use today. Only those pumps generally used to pump water from wells are described in the following paragraphs.

Table 6-2 Friction Loss for Water in Feet per 100 Feet (Schedule 40 Steel Pipe) (continued)

Flow gpm	16 in. V ft/s	16 in. h_f Frict.	18 in. V ft/s	18 in. h_f Frict.	20 in. V ft/s	20 in. h_f Frict.	24 in. V ft/s	24 in. h_f Frict.
1400	2.54	0.127						
1600	2.90	0.163						
1700	3.09	0.183						
1800	3.27	0.203	2.58	0.114				
1900	3.45	0.225	2.73	0.126				
2000	3.63	0.248	2.87	0.139	2.31	0.0812		
2500	4.51	0.377	3.59	0.211	2.89	0.123		
3000	5.45	0.535	4.30	0.297	3.46	0.174	2.39	0.070
3500	6.35	0.718	5.02	0.397	4.04	0.232	2.79	0.093
4000	7.26	0.921	5.74	0.511	4.62	0.298	3.19	0.120
4500	8.17	1.15	6.45	0.639	5.19	0.372	3.59	0.149
5000	9.08	1.41	7.17	0.781	5.77	0.455	3.99	0.181
6000	10.9	2.01	8.61	1.11	6.92	0.645	4.79	0.257
7000	12.7	2.69	10.0	1.49	8.08	0.862	5.59	0.343
8000	14.5	3.49	11.5	1.93	9.23	1.14	6.38	0.441
9000	16.3	4.38	12.9	2.42	10.39	1.39	7.18	0.551
10,000			14.3	2.97	11.5	1.70	7.98	0.671
11,000			15.8	3.57	12.7	2.05	8.78	0.810
12,000					13.8	2.44	9.58	0.959
13,000					15.0	2.86	10.4	1.42
14,000					16.2	3.29	11.2	1.29
15,000							12	1.48
16,000							12	1.67
17,000							13.6	1.88
18,000							14.4	2.10
19,000							15.2	2.33

Note: The table shows average values of pipe friction for new pipe. For commercial installations it is recommended that 15 percent be added to these values because no allowance for aging of pipe is included.

Centrifugal Pump

A centrifugal pump uses centrifugal force to move a liquid through a change in elevation or against a total dynamic head. The pump consists of a suction nozzle, an impeller eye, an impeller (rotating element), a volute or diffuser, and a discharge nozzle. As fluid is drawn in through the suction nozzle to the impeller eye, it is given a radial motion to cause a high velocity, developed by rotation of the impeller. The fluid is thrown from the outer tips of the impeller by centrifugal force into the volute or diffuser and on into the discharge line.

In both volute and diffuser types of centrifugal pumps, velocity head and, consequently, pressure are developed entirely by centrifugal force. In the volute-type pump (Figure 6-2), the impeller discharges into a gradually expanding case. This volute is designed to efficiently change part of the velocity head of the fluid leaving the impeller to pressure head. In the diffuser-type pump (Figure 6-3), the impeller is surrounded by progressively expanding passages of stationary guide vanes. The

Table 6-3 Equivalent Length of New Straight Pipe for Valves and Fittings for Turbulent Flow Only

Fittings			1/4	3/8	1/2	3/4	1	1 1/4	1 1/2	2	2 1/2	3	4	5	6	8	10	12	14	16	18	20	24
Regular 90° Ell	Screwed	Steel	2.3	3.1	3.6	4.4	5.2	6.6	7.4	8.5	9.3	11	13	—	—	—	—	—	—	—	—	—	—
		C.I	—	—	—	—	—	—	—	—	—	9.0	11	—	—	—	—	—	—	—	—	—	—
	Flanged	Steel	—	—	.92	1.2	1.6	2.1	2.4	3.1	3.6	4.4	5.9	7.3	8.9	12	14	17	18	21	23	25	30
		C.I.	—	—	—	—	—	—	—	—	—	3.6	4.8	—	7.2	9.8	12	15	17	19	22	24	28
Long Radius 90° Ell	Screwed	Steel	1.5	2.0	2.2	2.3	2.7	3.2	3.4	3.6	3.6	4.0	4.6	—	—	—	—	—	—	—	—	—	—
		C.I.	—	—	—	—	—	—	—	—	—	3.3	3.7	—	—	—	—	—	—	—	—	—	—
	Flanged	Steel	—	—	1.1	1.3	1.6	2.0	2.3	2.7	2.9	3.4	4.2	5.0	5.7	7.0	8.0	9.0	9.4	10	11	12	14
		C.I.	—	—	—	—	—	—	—	—	—	2.8	3.4	—	4.7	5.7	6.8	7.8	8.6	9.6	11	11	13
Regular 45° Ell	Screwed	Steel	.34	.52	.71	.92	1.3	1.7	2.1	2.7	3.2	4.0	5.5	—	—	—	—	—	—	—	—	—	—
		C.I.	—	—	—	—	—	—	—	—	—	3.3	4.5	—	—	—	—	—	—	—	—	—	—
	Flanged	Steel	—	—	.45	.59	.81	1.1	1.3	1.7	2.0	2.6	3.5	4.5	5.6	7.7	9.0	11	13	15	16	18	22
		C.I.	—	—	—	—	—	—	—	—	—	2.1	2.9	—	4.5	6.3	8.1	9.7	12	13	15	17	20
Tee-Line Flow	Screwed	Steel	.79	1.2	1.7	2.4	3.2	4.6	5.6	7.7	9.3	12	17	—	—	—	—	—	—	—	—	—	—
		C.I.	—	—	—	—	—	—	—	—	—	9.9	14	—	—	—	—	—	—	—	—	—	—
	Flanged	Steel	—	—	.69	.82	1.0	1.3	1.5	1.8	1.9	2.2	2.8	3.3	3.8	4.7	5.2	6.0	6.4	7.2	7.6	8.2	9.6
		C.I.	—	—	—	—	—	—	—	—	—	1.9	2.2	—	3.1	3.9	4.6	5.2	5.9	6.5	7.2	7.7	8.8
Tee-Branch Flow	Screwed	Steel	2.4	3.5	4.2	5.3	6.6	8.7	9.9	12	13	17	21	—	—	—	—	—	—	—	—	—	—
		C.I.	—	—	—	—	—	—	—	—	—	14	17	—	—	—	—	—	—	—	—	—	—
	Flanged	Steel	—	—	2.0	2.6	3.3	4.4	5.2	6.6	7.5	9.4	12	15	18	24	30	34	37	43	47	52	62
		C.I.	—	—	—	—	—	—	—	—	—	7.7	10	—	15	20	25	30	35	39	44	49	57
180° Return Bend	Screwed	Steel	2.3	3.1	3.6	4.4	5.2	6.6	7.4	8.5	9.3	11	13	—	—	—	—	—	—	—	—	—	—
	Reg. Flanged	Steel	—	—	.92	1.2	1.6	2.1	2.4	3.1	3.6	4.4	5.9	7.3	8.9	12	14	17	18	21	23	25	30
		C.I.	—	—	—	—	—	—	—	—	—	9.0	11	—	7.2	9.8	12	15	17	19	22	24	28
	Long Rad. Flanged	Steel	—	—	1.1	1.3	1.6	2.0	2.3	2.7	2.9	3.4	4.2	5.0	5.7	7.0	8.0	9.0	9.4	10	11	12	14
		C.I.	—	—	—	—	—	—	—	—	—	2.8	3.4	—	4.7	5.7	6.8	7.8	8.6	9.6	11	11	13

Table continues on next page.

WELL PUMPS AND PUMPING   83

Table 6-3 Equivalent Length of New Straight Pipe for Valves and Fittings for Turbulent Flow Only (continued)

Fittings			Pipe Size in.																			
			1/4	3/8	1/2	3/4	1	1 1/4	1 1/2	2	2 1/2	3	4	5	6	8	10	12	14	16	18	20	24
Globe Valve	Screwed	Steel	21	22	22	24	29	37	42	54	62	79	110	—	—	—	—	—	—	—	—	—	—
		C.I.	—	—	—	—	—	—	—	—	—	65	86	—	—	—	—	—	—	—	—	—	—
	Flanged	Steel	—	—	38	40	45	54	59	70	77	94	120	150	190	260	310	390	—	—	—	—	—
		C.I.	—	—	—	—	—	—	—	—	—	77	99	—	150	210	270	330	—	—	—	—	—
Gate Valve	Screwed	Steel	.32	.45	.56	.67	.84	1.1	1.2	1.5	1.7	1.9	2.5	—	—	—	—	—	—	—	—	—	—
		C.I.	—	—	—	—	—	—	—	—	—	1.6	2.0	—	—	—	—	—	—	—	—	—	—
	Flanged	Steel	—	—	—	—	—	—	—	2.6	2.7	2.8	2.9	3.1	3.2	3.2	3.2	3.2	3.2	3.2	3.2	3.2	3.2
		C.I.	—	—	—	—	—	—	—	—	—	2.3	2.4	—	2.6	2.7	2.8	2.9	2.9	3.0	3.0	3.0	3.0
Angle Valve	Screwed	Steel	12.8	15	15	15	17	18	18	18	18	18	18	—	—	—	—	—	—	—	—	—	—
		C.I.	—	—	—	—	—	—	—	—	—	15	15	—	—	—	—	—	—	—	—	—	—
	Flanged	Steel	—	—	15	15	17	18	18	21	22	28	38	50	63	90	120	140	160	190	210	240	300
		C.I.	—	—	—	—	—	—	—	—	—	23	31	—	52	74	98	120	150	170	200	230	280
Swing Check Valve	Screwed	Steel	7.2	7.3	8.0	8.8	11	13	15	19	22	27	38	—	—	—	—	—	—	—	—	—	—
		C.I.	—	—	—	—	—	—	—	—	—	22	31	—	—	—	—	—	—	—	—	—	—
	Flanged	Steel	—	—	3.8	5.3	7.2	10	12	17	21	27	38	50	63	90	120	140	—	—	—	—	—
		C.I.	—	—	—	—	—	—	—	—	—	22	31	—	52	74	98	120	—	—	—	—	—
Coupling or Union	Screwed	Steel	.14	.18	.21	.24	.29	.36	.39	.45	.47	.53	.65	—	—	—	—	—	—	—	—	—	—
		C.I.	—	—	—	—	—	—	—	—	—	.44	.52	—	—	—	—	—	—	—	—	—	—
Bell-Mouth Inlet		Steel	.04	.07	.10	.13	.18	.26	.31	.43	.52	.67	.95	1.3	1.6	2.3	2.9	3.5	4.0	4.7	5.3	6.1	7.6
		C.I.	—	—	—	—	—	—	—	—	—	.55	.77	—	1.3	1.9	2.4	3.0	3.6	4.3	5.0	5.7	7.0
Square-Mouth Inlet		Steel	.44	.68	.96	1.3	1.8	2.6	3.1	4.3	5.2	6.7	9.5	13	16	23	29	35	40	47	53	61	76
		C.I.	—	—	—	—	—	—	—	—	—	5.5	7.7	—	13	19	24	30	36	43	50	57	70
Re-entrant Pipe		Steel	.88	1.4	1.9	2.6	3.6	5.1	6.2	8.5	10	13	19	25	32	45	58	70	80	95	110	120	150
		C.I.	—	—	—	—	—	—	—	—	—	11	15	—	26	37	49	61	73	86	100	110	140
Sudden Enlargement			$h = \dfrac{(V_1 - V_2)^2}{2g}$ feet of liquid; if $V_2 = 0$ $h = \dfrac{V_1^2}{2g}$ feet of liquid																				

diffuser pump does a more complete job of converting velocity head to pressure and, consequently, it is more efficient than the volute type.

The centrifugal force of the pump moves liquid to the pump's outer case by rotation of the impeller, creating an area of low pressure at the eye of the impeller. If this pressure is lower than atmospheric pressure, the water will be pushed into the space between the blades of the impeller and a pumping action will be developed.

When this type of pump is used in pumping wells, the total suction lift or well depth below the pump centerline that can be pumped is regulated by the atmospheric pressure. If atmospheric pressure is considered to be 14.7 psi and a perfect vacuum is present, this pressure could support a column of water equal in length to 14.7 × 2.307 = 33.9 ft; therefore, if the centrifugal pump could produce a perfect vacuum, the total theoretical lift would be 33.9 ft. Because it is impossible to produce a perfect vacuum at sea level with a pump, the practical suction height varies from 60 percent to 85 percent of the theoretical possible distance, depending on the efficiency of the installation.

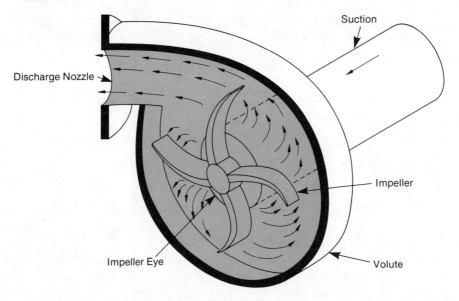

Figure 6-2 Volute-type centrifugal pump.

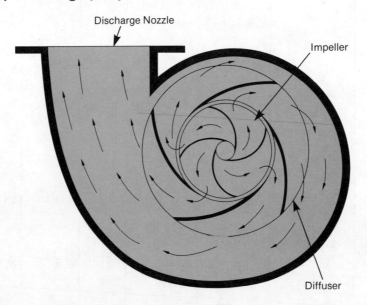

Figure 6-3 Diffuser-type centrifugal pump.

Table 6-4 Maximum Practical Suction Lift, in Feet, for Single-Stage Centrifugal Pump

Elevation Above Sea Level ft	Maximum Practical Suction Lift ft Temperature of Water, °F										
	60	70	80	90	100	110	120	130	140	150	160
0	22	20	17	15	13	11	8	6	4	2	0
2000	19	17	15	13	11	8	6	4	2	0	
4000	17	15	13	11	8	6	4	2	0		
6000	15	13	11	8	6	4	2	0			
8000	13	11	9	6	4	2	0				
10,000	11	9	7	4	2	0					

Table 6-4 presents values of the practical suction lift, in feet, for a single-stage centrifugal pump operating at different elevations. Since the use of a single-stage centrifugal pump is restricted to shallow wells (less than 20-ft [6-m] depths), it has been necessary to develop other types of equipment to pump water from deeper wells.

Deep-Well Turbine Pump

In deep-well pumping, the reciprocating pump was superseded by the centrifugal pump as increased volumes of water became necessary. Diminishing water tables, excessive costs of developing the deep pits that are used to place centrifugal pumps within reasonable suction lifts, and difficulty in providing efficient drivers fostered development of the deep-well turbine pump. Actually, the deep-well turbine pump is not truly a turbine, but rather a combination of several stages of centrifugal impellers connected in series to a common shaft.

The deep-well turbine pump (Figure 6-4) consists of a prime mover; a suitable shaft and bearings connecting the power source on the surface to impellers located under the well water; a series of impellers mounted in the bowl assembly at the lower end of the column that produces the required pressure head; and a discharge column pipe that channels water to the surface and acts as a housing and guide for the bearings and shaft assembly. This type of pump was designed for capacities as low as 10 or 15 gpm (40 or 60 L/min) and as high as 25,000 gpm (95,000 L/min) or more, and for heads up to 1000 ft (300 m). Most applications involve smaller capacities.

The pump illustrated in Figure 6-4 is of a three-stage design. Each stage consists of a bowl, impeller, and diffuser manufactured as a standard unit. The number of bowls required for a particular installation depends on the dynamic head. The head against which the pump will be required to operate will determine the number of stages that must be provided. For large capacities, more than one pump may be needed. The capacity of the pumps used for bored wells is limited by the physical size of the well casing and by the rate at which water can be drawn without lowering its level to a point of insufficient pump submergence.

Submersible Pump

A submersible pump is actually a turbine pump in which the motors are close-coupled beneath the bowls of the pumping unit, and the entire unit is installed under water. This type of construction eliminates the need for a surface motor, long drive shaft, shaft bearings, and lubrication system, which are common to the conventional turbine pump. Submersible pump motors are cooled by water flowing vertically past the motor to the pump intake. The motor is usually longer and of smaller diameter than

86 GROUNDWATER

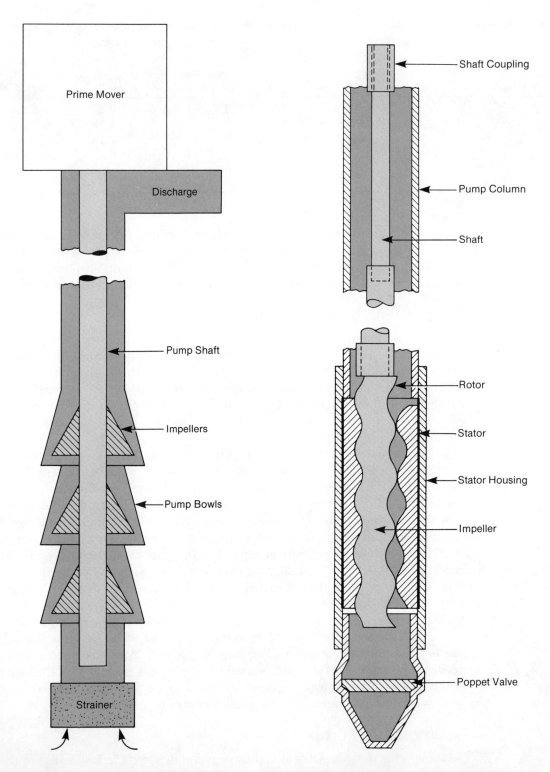

Figure 6-4 Vertical deep-well turbine pump.

Figure 6-5 Rotary-displacement pump.

a surface motor of the same horsepower. When a large-capacity submersible pump is contemplated, the manufacturer should be consulted for specific design and installation recommendations.

The purchase and installation costs for a submersible pump may be higher or lower than for a conventional pump, depending on setting depth, required head and capacity, water corrosivity, and other factors. Operating costs may also be higher or lower, based on motor efficiency, column bearing, hydraulic losses, cable losses, setting depth, and similar factors. A thorough analysis of all factors should be performed to compare surface and submersible motor-driven deep-well pumps for a specific installation.

Some inherent advantages of submersible pumps include

- use in crooked well casings that are unsuited for other types of pumps;
- use in wells subject to flooding; the wells can be completely sealed;
- minimization of surface equipment; and
- silent operation.

Submersible pumps are especially useful for high-head, low-capacity applications, such as domestic water supply.

Propeller and Mixed-Flow Pumps

Propeller, axial flow, mixed flow, screw, and spiral type pumps have found limited use in the production of shallow wells. These designs have open impellers, similar to a ship's propeller, and are adaptable to installations where flow is generally greater than 300 gpm (1100 L/min), but where heads are under 40 ft (10 m).

Rotary Pump

A rotary pump combines the positive-discharge characteristics of the reciprocating pump with the constant nonvarying discharge feature of the centrifugal pump. Although a rotary pump uses a rotating element and appears similar to a centrifugal pump, pumping is performed through a positive-displacement action. Direct water pressure is built up in the pump case by means of specially designed runners that squeeze the water between them as they rotate.

A well-designed rotary pump will create a relatively high vacuum, comparable in magnitude to that created by a centrifugal pump. However, rotary pumps are usually not as efficient as centrifugal pumps. Unless rotary pumps are well designed and constructed of the best material, they will wear much faster. Still, the rotary-type pump is widely used.

Rotary-Displacement Pump

A rotary-displacement pump is designed especially for relatively low capacities and for installation in cased wells that are 4 and 6 in. (100 and 150 mm) or larger. Flow in a rotary-displacement pump results from the displacement of a piston in a cylinder of indefinite length. Figure 6-5 illustrates the pumping element, which consists of a main body made up of a stator and rotor, both of helical form, and the drive-shaft assembly. The helices are really worm threads, the stator has a double thread, and the rotor offers a corresponding single thread. This pump is of the positive-displacement type.

As the rotor rolls on the inner surface of the stator, liquid is squeezed ahead by the rolling action, with minimum turbulence. The rotor is made of heat-treated stainless steel that has a hard, chrome surface to resist corrosion and abrasion. A one-piece

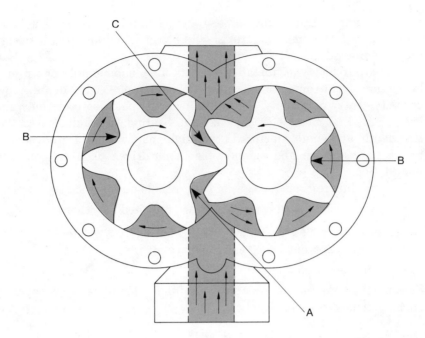

Figure 6-6 Rotary-gear pump.

bronze strainer with a rubber-seated foot valve keeps the column full of water, and no prelubrication is necessary. The stator is of cutless rubber and is highly resistant to abrasion. Grit momentarily depressed into the rubber by the rotors is washed away by water, when the rotor is released.

Rotary-Gear Pump

A rotary-gear pump consists of two moving parts—the pumping gears (Figure 6-6). These gears rotate in an accurately fitted case with close tolerance that ensures efficiency. The teeth of the pumping gears move away from each other and pass the inlet port at point A (see Figure 6-6). This produces a partial vacuum by withdrawing air into the pump, where it is carried between the teeth of the pumping gears around both sides of the pump case at point B. The action of the teeth meshing at point C results in a condition similar to a valve forming a seal that forces the water into the discharge line.

Water flow is continuous and steady in a rotary-gear pump. The quantity of liquid pumped per hour is determined by the size of the pump and the rotational speed of the pump shaft. All internal parts, including the bearings, are lubricated by the flow of water. The rotary-gear pump is suitable for suction of 22–25 ft (7–8 m).

Reciprocating Pump

The oldest type of deep-well pump is the plunger-type, or reciprocating, pump (Figure 6-7), which consists of a belt- or gear-driven head located above the highest water level in the well. A pulley drives a pinion shaft, and through suitable gearing, the plunger rod is made to work up and down in the well. The prime mover is connected to the working, or pumping, barrel by means of pump rods.

The working barrel may be single or double acting. In the single-acting type, a check valve is located at the bottom of the cylinder and a similar valve is located in the plunger. The water flows into the working barrel through the check valve while

WELL PUMPS AND PUMPING 89

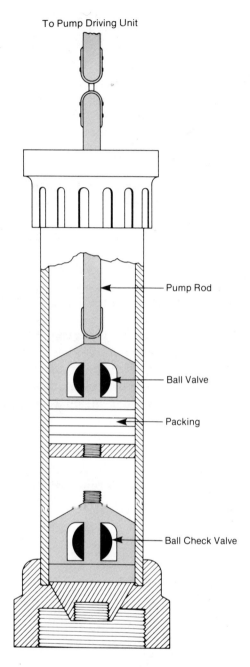

Figure 6-7 Plunger-type pump.

the plunger is making its upstroke. On the downstroke, this water is held in the working barrel by the foot valve, and the plunger descends to the bottom of the barrel while the water passes through the valve in the plunger. On the next upstroke, the valve in the plunger closes, and the water above it is raised into the discharge pipe. At the same time, the foot valve opens and the cylinder again fills with water.

In small-diameter wells, a check may be set in the casing below the water level and the plunger sized to the casing, which then becomes the working barrel cylinder. In this case, the rods work through a stuffing box at the top of the casing, and water is discharged out of a side-opening tee.

Double-acting cylinders are designed to discharge water on each downstroke as well as on each upstroke of the working head. Double-acting pumps are capable of producing about 60 percent greater flow than pumps equipped with single-acting working barrels.

The capacity of this type of pump depends on the displacement of the liquid in the working barrel and the number of strokes per minute of operation. The pump is theoretically suitable for pumping wells of any depth, such depth being dictated by strength of material, power source, and economics.

Impulse Pumps

Impulse pumps produce a pumping action by the direct application of pressurized air or water. Thus, the two types of impulse pumps—airlift and jet—will be discussed in the following paragraphs.

Airlift pumps. Airlift pumps have capacities up to 2000 gpm (8000 L/min) and head to 1000 ft (300 m). They are used in water wells, especially those containing sandy or corrosive fluids. The pump consists of a vertical pipe submerged in the well and an air-supply tube, through which compressed air is fed to the pipe at a considerable distance below the static water level. When air is introduced into the pipe, the resultant mixture of air bubbles and liquid, being lighter in weight than the liquid outside the pipe, rises in the pipe. As air is continuously introduced at the bottom, a continuous flow of mixed water and air emerges at the top of the pipe, with new liquid from the well entering the pipe at the bottom (Figure 6-8).

Because the only head-producing mechanism in this type of pump is the difference in specific weight of the water–air mixture inside the pipe and the water outside the pipe, the head that can be obtained from an airlift pump depends on the distance between water level in the casing and location of air introduction. If head H is measured from the discharge pipe to water level and submergence S is measured from water level to introduction of air, the ratio H/S is approximately 1 for most applications, reaching 3 for high heads (and low flows) and going as low as 0.4 for low heads (and high flows).

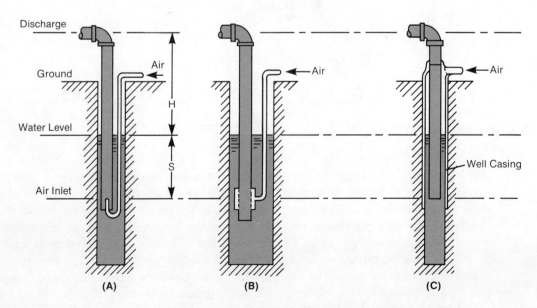

Figure 6-8 Airlift pump: (A) bottom inlet; (B) side inlet; (C) casing inlet.

Table 6-5 Air Requirements* for Airlift Pumps

H\\H/S†	3	2	1	0.67	0.4
20				0.22	0.15
50				0.3	0.2
100				0.4	0.3
150			0.7	0.5	
200			0.8	0.6	
300		2.1	1.0		
400		2.3	1.2		
500	3.25	2.6	1.4		
650	3.75	3.0	2.1		
800	4.2	3.5			
950	4.7	3.9			

*Number of cubic feet per minute of free air required to pump 1 gpm of water.
†H = head; S = submergence, in feet.

The capacity of water pumped depends on the amount of air supplied. The pumping capacity increases with increased air supplied up to an optimum amount of air. However, since the discharge is a mixture of liquid and air, the introduction of more air than the optimum amount will actually decrease the net quantity of water delivered. Table 6-5 indicates approximate amounts, in cubic feet per minute, of free air required to pump 1 gpm of water against the heads of relative submergence shown.

The advantages of airlift pumps include no moving parts, usability for corrosive and erosive fluids, gentle action (has been used to remove sand from buried undersea objects), operation on air (can be used in explosive atmospheres), and ability to be placed into wells of irregular shape where regular deep-well pumps cannot fit. The disadvantages of airlift pumps include low efficiency (less than 40 percent) and the need for very large submergences compared to conventional pumps.

Jet pumps. A second type of impulse pump is the jet type shown in Figure 6-9. Water is forced down through a nozzle, forming a jet, and is discharged into the throat of a venturi diffuser at high velocity. Water, under pressure, is delivered to this jet, or ejector, nozzle, which converts the pressure into velocity. Because of the pressure, water discharges into the diffuser at a high velocity, causing a lower-pressure area at this point. Water then flows in from the well and mixes in the diffuser with the driving water. While passing through the tube, most of this high velocity is transformed into pressure, and this delivers both the driving water and the water draw from the well to a high elevation.

The efficiency of this type of pump is rather low, about 25 to 30 percent, but the pump has no moving parts submerged in the well and is quite appealing where high capacities are not required. A jet pump is best suited for operations that offer a lift of 25 ft (8 m) or more and for required capacities less than 50 gpm. This pump is often used for 120-ft (40-m) lifts. In many instances, it is used on wells as deep as 150 ft (45 m). On deeper wells, jet-pump efficiency becomes very low and another type of pump is usually applied. Jet pumps are light and can pump very muddy or sand-loaded water. Centrifugal–jet pump combinations have been used to pump wells as deep as 400 ft (120 m); centrifugal pumps alone usually are limited—from 20- to 25-ft (6- to 8-m) well depths.

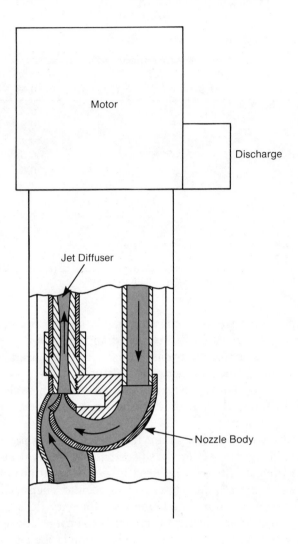

Figure 6-9 Jet-type deep-well pump.

OPERATING CONDITIONS

Continuous operation of a pump is generally preferable to intermittent operation, but varying water demand usually requires some combination of off and on time. For improved well performance and pump life, system components and storage capacity should be designed to minimize the number of pump starts and stops per day. While pump starts should be minimized, starting a pump several times per day or even more than once per hour adds only slightly to power consumption and normally gives acceptable life to pumps, motors, and controls.

To allow for continuous running of pumps, installations sometimes include a pressure-regulating valve or variable-speed drive that can match the pump output with the system demand. Such systems usually run at very low efficiency during low-demand periods. The overall cost of equipment and operation should be thoroughly analyzed before adopting such a system.

In order to achieve the lowest-cost operation, a system must run its pump and motor or engine in the best efficiency ranges. Proper initial selection of system components can assure this, but changing conditions sometimes justify altering or reselecting components to maintain economical operation.

For each type of pump and prime mover, the operating conditions must be checked against manufacturers' application information to ensure reliable operation. These operating conditions include the ambient air and water temperature ranges, pressures, flow, corrosive and abrasive factors, power supply variation, duty cycle, and protective devices.

PUMP SELECTION

The proper selection of a pump for any application primarily depends on economics. The type of pump selected should give the best and most economical service over a prescribed number of years, when pumping under specific conditions. Before selecting a pump, it is necessary to investigate different types of pumping arrangements and, in each instance, to tabulate the initial cost, cost of installation, cost of operation, cost of maintenance, and expected equipment life. From this tabulation, it is fairly simple to select the combination that offers minimum investment and operational costs and yet fulfills system requirements.

Factors in Pump Selection

Some factors that must be considered before selecting a particular pump include capacity; depth of well and pumping level; inside diameter of well; condition of bore (straight or crooked); abrasive properties; total head; type of available power; and costs.

After a well has been tested for yield, the requirements placed on it are dictated by demand. For some wells, it is advisable to keep a constant drawdown to prevent production of sudden changes in water quality or of sand. Well-depth and pumping-level characteristics have been considered previously and are not repeated here, but the size and condition of the well bore are very important. For example, if a well is too crooked for efficient operation of a deep-well turbine pump and use of a plunger-type pump is not desirable, it may be more economical in the long run to install a submersible pump, even though initial installation might be more expensive. A chemical analysis should be run on the water to be pumped. A pump built with special materials may be needed to keep corrosion cost at a minimum. Total head must be determined because it is necessary not only to lift or push the water to ground-surface elevation, but also to pump this water to storage or to the consumer. Sufficient head needed to perform both tasks determines the motivating horsepower required, which will, in turn, lead to accurate cost estimation.

Measuring pump performance. The cost of energy is one of the principal expenses incurred in pump operations. Therefore, pumps should be monitored to ensure that they are operating at or near peak efficiency. The following three factors must be measured when checking pump efficiency: total head, input horsepower, and quantity of water pumped.

Estimating total pumping head. The total dynamic head against which a pump operates includes the vertical distance from the water level in the well, while pumping, to the center of the free-flowing discharge, plus all losses in the line between the point of entry of the water and the point of discharge. Losses in pipe can be obtained from Table 6-2, and pump-column losses are available from pump manufacturers' catalogs.

Estimating pump input horsepower. The water horsepower required to pump water can be determined by the following equation:

$$\text{whp} = \frac{\text{gpm} \times H_T}{3960} \qquad \text{(Eq 6-2)}$$

Where:

> gpm = the flow rate, in gallons per minute
> H_T = total head, in feet.

Measuring pump efficiency. The horsepower calculated using Eq 6-2 is that necessary if all equipment (pumps, prime mover, and the like) were operating at 100 percent efficiency. Since 100 percent efficiency cannot be attained, the brake horsepower, or horsepower necessary at the pump shaft, is used. Brake horsepower can be obtained from manufacturers' data tables or performance curves. The pump efficiency E_p can be calculated from Eq 6-3

$$E_p = \frac{\text{whp}}{\text{bhp}} \qquad \text{(Eq 6-3)}$$

Since the prime mover is not 100 percent efficient, the total horsepower required to operate the system is motor horsepower

$$\text{mhp} = \frac{\text{bhp}}{E_M} = \frac{\text{gpm} \times H_T}{3960 \times E_p \times E_M} \qquad \text{(Eq 6-4)}$$

Where:

> E_M = the efficiency of prime mover.

The efficiency of an electric motor as a prime mover is usually between 60 and 95 percent, depending on size and type, but an exact value can be obtained from manufacturers' information.

The overall efficiency of a pump system depends on many factors, such as specific speed, relative size, service materials, and physical characteristics of fluid. Large centrifugal pumps have developed more than 92 percent efficiency, whereas the efficiency of smaller pumps may, in some instances, be only 20 or 25 percent.

To determine the overall efficiency of a pumping system, consider the efficiency of the pump, prime mover, and drive. The overall efficiency E then can be determined

$$E = E_p \times E_M \times E_D \qquad \text{(Eq 6-5)}$$

Where:

> E_p = efficiency of the pump
> E_M = efficiency of the prime mover
> E_D = efficiency of the drive.

$$\text{power required} = \frac{\text{theoretical power required}}{E} \qquad \text{(Eq 6-6)}$$

If an electric motor is used to drive the pump, the actual power required will be equal to the theoretical power times the results of Eq 6-6. To determine the cost of operation, one must convert horsepower to watts. One horsepower is equivalent to 746 W and 1 kW is equal to 1000 W; thus,

$$\text{kW demand} = \frac{\text{gpm} \times H_T \times 0.746}{3960 \times E} \qquad \text{(Eq 6-7)}$$

Power is usually sold in units of kilowatt-hours, and when Eq 6-7 is multiplied by number of hours used, it will give

$$\text{kW·h} = \frac{\text{gpm} \times H_T \times 0.746 \times \text{hours}}{3960 \times E} \quad \text{(Eq 6-8)}$$

Total costs can be determined by multiplying Eq 6-8 by cost per kilowatt-hour.

$$\text{total cost} = \frac{\text{gpm} \times H_T \times 0.746 \times \text{hours} \times \text{cost/kW·h}}{3960 \times E} \quad \text{(Eq 6-9)}$$

or

$$\text{power cost per hour} = \frac{\text{gpm} \times H_T \times 0.746 \times \text{cost/kW·h}}{3960 \times E} \quad \text{(Eq 6-10)}$$

If a different type of power is used, cost per hour can be calculated in a similar manner. Standby equipment also should be provided.

In general, pumps are driven by direct connection to prime movers or through use of right-angle drives or belts. Electric motors and gasoline or diesel engines usually are used as prime movers for water-well pumps.

Measuring quantity of water pumped. Several methods of measuring the quantity of water pumped are available. These methods are described in chapter 11.

Operational Limits of Pumping Units

The operational limits of ordinary pumping units have been stated earlier in this chapter. In summary, a single-stage centrifugal-type pump offers satisfactory and economical service, and sizes are available for practically any desired capacity. However, application is restricted, because the suction lift must not be more than about 22 ft (7 m), depending on elevation and temperature (see Table 6-4).

In a well that is clean and free of sand or grit, a rotary-type pump may perform as satisfactorily as a centrifugal pump, but rotary units are applicable only for operations that present low flow rates. Suction-lift specifications for rotary-type pumps are the same as for centrifugal pumps.

A centrifugal and jet, or ejector, combination pump can be used to produce low rates of flow and suction lifts to 120 ft (37 m). In deeper wells, jets sometimes are used in combination with positive-displacement pumps.

For capacities exceeding a few gallons per minute (10 or 11 L/min) and settings deeper than 30 ft (9 m), a multiple-stage deep-well turbine pump that is driven directly by a submersible motor or through shafting by a surface-mounted motor or engine is usually selected. The choice between submersible and surface-driven turbines can be made based on analysis of initial costs and operating costs, acceptability of aboveground structures and noise, likelihood of vandalism, well conditions, available power, and other factors specific to a particular installation.

For all but positive-displacement pumps, the discharge head increases as the rate of flow or capacity decreases, and the discharge head decreases as the rate of flow or capacity increases; therefore, if constant discharge under a varying head is to be maintained by a centrifugal pump, a variable-speed drive must be used. No problem is encountered when a positive-displacement pump is used because the capacity depends on the speed of the pump. The pressure that can be developed by a plunger-type pump is limited only by the size of the power unit and strength of materials.

ELECTRIC MOTOR SELECTION

Electric motors are usually selected according to National Electrical Manufacturers Association* standards, which include requirements for enclosures and cooling methods. An electrical specialist should be consulted for advice and assistance in selecting electric motors.

PUMP INSTALLATION

Proper pump installation is necessary in order to obtain maximum pump efficiency, minimize maintenance, and prolong the life of associated piping. This section covers installation of pumps and associated piping.

Aboveground Installation

A good foundation, preferably concrete, should be available for pump placement. Foundation bolts should be placed according to the dimensions that are usually furnished by pump manufacturers. The pump should be placed on the foundation so that it is easily accessible for regular inspection during operation. Room should be provided for use of a crane, hoist, or tackle. Pits in which pumps are placed should be safeguarded against flooding.

Alignment. Pumps should be properly aligned by leveling the base; the pump and driving unit can be brought into exact alignment with shims. Most pump bases, no matter how rugged, will spring and twist to some degree during shipment. Consequently, alignment is crucial when the unit is being installed.

Piping should line up naturally and be supported independently of the pump to eliminate strain on the pump casing; it should not be forced into place with flange bolts. After the piping has been installed, alignment should be rechecked. On unusually long discharge lines, a packed slip joint should be installed to compensate for elongation of pipe that might result from pressure or temperature changes.

Piping. To protect the pump, a gate valve and check valve should be installed in the discharge pipe close to the pump. The check valve should be placed between the pump and a gate valve. If pipe connections are used on the discharge end of the pump to increase the size of discharge piping, the connections should be placed between a check valve and the pump. The selection of the discharge piping should be made with due reference to expected friction losses.

After the piping has been completed, alignment should be checked again using a straight edge and thickness gauge. The manufacturer's installation checklist and adjustment directions should be closely followed and double checked before applying power to the pump unit. When pumping units have been aligned before piping is completed, piping strains that develop are probably the cause of any misalignments. Changes should be made accordingly. If stuffing boxes are adjusted properly and the pump and drives are aligned properly, the unit can usually be operated by hand with ease.

---

*National Electrical Manufacturers Association, 2101 L St. N.W., Washington, DC 20037.

Deep-Well Installation

A deep-well pump driven by either a submersible motor or an aboveground driver must be installed in accordance with the manufacturer's instructions. It is important that the pump be sized and set so that it will never run for even a few minutes at no delivery, which could occur if excessive head is present or if the pumping level is lowered to the intake area. Running with little or no delivery is likely to damage the pump bearings and cause overheating failure of a submersible motor. If the well drawdown or the delivery system could cause the pump to run at little or no delivery, protection should be provided to ensure flow. Such protection could be in the form of a flow switch or well-level switch that would shut off the pump or sound an alarm if flow or water level dropped below a safe minimum. The minimum water level above the pump intake should always be kept greater than the required suction head (NPSH) specified by the manufacturer.

The materials used in the pump and delivery system must be resistant to significant corrosion caused by normal water conditions in the well or any periodic chemical cleaning operations performed with the pump in place. Additionally, the well must be properly designed and developed prior to installation of the production pump to minimize sand pumping. Most pump and submersible-motor warranties do not cover failures from abrasive damage and corrosion.

Check valve. Unless some unusual requirement prevents it, a check valve should always be installed within 25 ft (8 m) of a deep-well surface-driven or submersible pump. This prevents problems that occur when water in the delivery pipe is allowed to flow back into the well each time the pump is turned off. These problems include

- backwashing, which can disturb the stabilized particles located outside screens and perforations, often increasing sand and turbidity in the well.

- backflow, which may spin the pump at high speed in reverse, causing damage or shortened life. This problem will not occur if the pump is designed to withstand high speed or if it is equipped with a device to prevent backspin. Attempting to restart a pump during backspin decreases bearing life and may cause tripping of protective devices with prolonged starting current.

- refilling of the delivery pipe at each start, which wastes power.

- operation of the pump with upward thrust until water height in the delivery pipe is adequate to create downward thrust. This increases wear and can cause problems, unless the pump is designed for repeated up-thrust.

- lack of a check valve near the pump. Aboveground check valves and shutoff valves, which are often required, can create a vacuum in a section of the delivery pipe after the pump turns off. This occurs because atmospheric pressure can only support water in the pipe to less than 34 ft (10 m) above the level in the well. When this evacuated section refills on starting, the moving water striking the stationary water at the closed valve creates a severe hydraulic shock (water hammer), which can cause pipe, valve, pump, or motor failure.

Addtional check valves may be required, depending on setting depth, valve rating, and aboveground equipment. A check valve in the delivery pipe of a submersible pump will hold the pipe full of water if the pump is removed from the well. This causes a major increase in weight and a flood when each pipe joint is removed. For this reason, special check valves are often used in which a small replaceable plug can be broken off to create a drain by dropping a weight down the well before pulling the pump.

Some other major considerations, usually covered in detail by the manufacturer's installation data for surface-driven deep-well pumps, include column pipe assembly, bearings and shafting, lubrication, alignment, mounting and aligning of the aboveground drive, setting of the impeller position, and use of proper controls. Considerations for submersible pumps include a setting that prevents motor burial in sand or silt, water temperature and flow past the motor to provide proper cooling, use of cable and splices that meet the amperage and voltage requirements, pipe tightening to prevent unscrewing by motor starting torque, clamping of cable to delivery pipe, provision of proper controls and protections, and necessary checks before, during, and after installation.

AWWA MANUAL M21

Chapter 7

Common Pump Operating Problems

Pumps are subject to numerous operating problems that can cause serious problems in a water supply well. Common operating problems that are discussed in this section include breaking suction, screen incrustation, and sand pumping.

BREAKING SUCTION

A pump should not be permitted to operate at a rate or level at which it can break suction (lose prime in a pump). If a pump does break suction and pumping stops, the discharge must be reduced by partially closing the discharge gate valve until the pumping level in the well remains above the pump bowls. Closing the valve increases the head loss in the system, causing the pump to work against a greater total dynamic head and decreasing output in accordance with the pump's head-capacity characteristics.

A certain amount of power is wasted by this procedure. In most cases to regain a more efficient operation, the pump is set deeper (if conditions permit) or one bowl and impeller is removed from the pump assembly to change the operating characteristics of the pump.

Causes

A lower pumping level in a well that has previously operated satisfactorily may result from either of two causes

- the water table (nonpumping level) in the vicinity of the well may have dropped so that the pumping level was correspondingly lowered; or
- the intake portion of the well may have become clogged with incrusting material, so that greater drawdown had to be created to cause water to flow from the formation into the well at a given rate.

100 GROUNDWATER

Lowered water table. The water table in the vicinity of a well may recede seasonally or during long dry periods when recharge to the aquifer is at its minimum. An aquifer may be reduced if the stored groundwater is being gradually depleted by pumping. Finally, the successive installation of additional wells in an area with overlapping cones of depression can also cause the water table to recede. A receding water table will cause significant mutual interference, in which the overlapping cones of depression cause reduction in the water levels of the wells. Consequently, pumping levels will be lowered to a point that is lower than that found in a single operating well.

Figure 7-1 illustrates the operating problem that results from a drop in the water table caused by any of these occurrences. Curve 1 represents the relationship between well yield and pumping level. Curves 2 and 3 represent lower pumping levels caused by recessions of the water table. The drawdown in each case is the difference between the depth to water at zero discharge and any other point on one of these

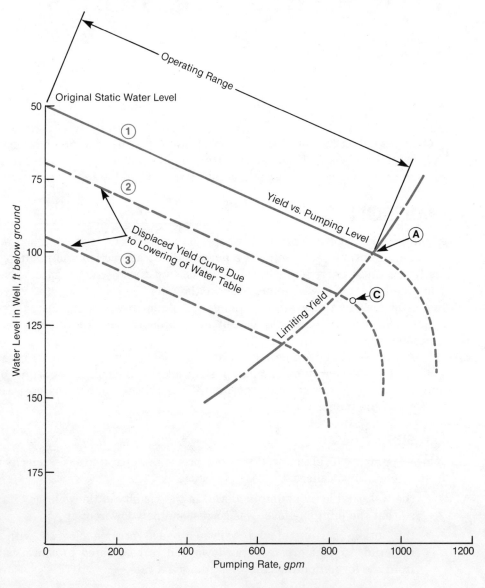

Figure 7-1 Operating problems resulting from a drop in the water table.

curves. The limiting yield is indicated as the condition where increase in yield ceases to be approximately proportional to increase in drawdown.

Assume that the pump was originally designed to operate at point A on curve 1. If a drop in water level changes the well performance to curve 2, the same pump may operate at about point C, which is undesirable. The situation can be improved only by cutting back the pumping rate to less than 800 gpm (3000 L/min).

Clogged intake. When a pump breaks suction, the water level fluctuates violently and produces a surging action in the well and in the water-bearing formation outside the well. In a well drawing from very soft sandstone or a screened well that may not have been completely stabilized during construction, the surging action may dislodge fine materials and cause sand pumping.

Also if suction is lost, air becomes entrained with the water, causing it to appear milky. The pulsation of the discharge from the pump, plus expansion of the entrained air bubbles, may cause disturbance in the distribution system. This can dislodge corrosion products or other incrustations from the inside of pipes, and complaints of dirty water will follow immediately.

Solutions

Water levels in all wells should be measured and recorded routinely, according to a program that fits the operation. A continuing record of both nonpumping and pumping levels should be maintained. If the pumping level recedes in any well, the cause should be determined. Timely adjustment of the pumping equipment should be made if danger of breaking suction becomes apparent.

SCREEN STOPPAGE—INCRUSTATION

Incrustation is the most common cause of decreasing well capacity. It is caused by clogging of the water-bearing formation just outside the well bore and clogging of the openings in the aquifer or well screen.

Incrustation is often a hard, brittle, cement-like deposit similar to the scale that forms in water pipes. Under certain conditions, it may be a soft, paste-like sludge or a gelatinous material.

The different forms of incrustation include precipitation of carbonates of calcium and magnesium or their sulfates; precipitation of iron and manganese compounds, primarily their hydroxides or hydrated oxides; slime produced by iron bacteria or other slime-forming organisms; deposition of soil materials, such as silt and clay, carried to the screen in suspension (not frequently encountered). Deposition of soil materials is most likely to occur where the screen openings are too small, where the well has been improperly developed and finished, or where the water-bearing formation contains an abnormal amount of these fine materials.

Water Quality

Water quality is critical to the occurrence of incrustation. The kinds and amounts of dissolved minerals and gases in natural waters determine the tendency to deposit some of the mineral matter as incrustation. The dissolved substances in groundwater are often present in a delicately balanced condition. When something happens to upset the balance, some minerals come out of solution and are deposited as solid materials.

Deposition of only a minute fraction of the minerals in water will, in time, cause serious clogging. For example, if 1 ppm of material drops out of the water per day from a 12-in. well pumping 500 gpm, a total deposit of 6 lb/24 h will result. At this rate, all voids in the sand 6 in. outside the screen would be completely filled in 220 days.

Preventing Incrustation

Thus far, a means of entirely preventing the incrustation of well screens has not been found. Certain things can be done, however, to delay incrustation and make it a less serious problem.

To prevent incrustation, the well bore itself should have the maximum possible inlet area in order to reduce the velocity of flow through the openings to a minimum. The length of the screen or penetration of the aquifer, or both, should be adequate, and the correct method of developing the formation surrounding the bore should be used.

Second, the pumping rate may be reduced, under some circumstances, and the pumping period increased. This procedure produces benefits to the extent that the drawdown and thus the velocity are decreased.

Third, periodic maintenance or cleaning of each well should be performed, based on local experience. Corrective measures should not be put off; this is particularly true in areas where incrustation problems are known to exist.

Treating Incrustation

Ideally, the occurrence of incrustation should be prevented or retarded. However, if incrustation does occur, effective removal procedures must be followed. There are two relatively successful removal methods in use today: pulling the incrusted screens, treating them to remove the incrustation, and resetting the screens; and treating the screens and the water-bearing formation directly around the screen with acid or other chemicals without pulling the screens.

Screen removal. Obviously, removing the screen is effective, but it does involve considerable labor. If the incrustation is very thick and hard, the force required to withdraw the screen occasionally results in breakage of the pulling pipe or damage to the screen. Some risk is also involved in replacing the screen after treatment. Moreover, it is often inconvenient to withdraw the screens, especially when the wells are inside buildings.

Treatment with acid. Before chemical treatment is applied to a well, the real reason for its decline in production should be studied. This analysis will usually reveal whether chemical treatment can be expected to be effective. The history of the well, its production record, variation in drawdown, and variation in static water level should be researched to obtain data to be used as a guide in proceeding with subsequent work.

Necessary conditions for acid treatment. Successful acid treatment of a screened well may be carried out under the following conditions:

- the well screen must be made of a metal that will not be damaged by the acid. It should be made of a single metal to avoid rapid galvanic corrosion.
- the incrustation should be analyzed to determine the proper procedure. A water analysis is also needed. Samples of incrustation taken from other wells in the area may be available for examination.
- access to the well must be provided. It is usually necessary to remove the pump from the well.
- nearby wells that are pumping from the same formation must be shut down during the treatment period. It is suggested that wells within 100 ft (30 m) be kept idle while acid is being used.

Hydrochloric acid. Calcium and magnesium carbonate incrustations are efficiently eliminated by introducing hydrochloric acid into the well, working it out through the screen openings and into the voids of the formation, and then removing the loosened material by pumping. Iron and manganese hydroxides and oxides are also quite soluble in hydrochloric acid, although the metals will precipitate out of an acid solution if the pH is above 3. To remove these compounds, the proper strength of acid must be maintained until it is pumped out of the well. As a means of keeping iron in solution, a stabilizer, such as Rochelle salt, may be added to the acid.

Even though iron and manganese incrustations are acid-soluble, acid treatment has at times failed to give good results. Some of these failures can probably be attributed to improper technique, but it appears that use of another method of chemical treatment is preferable under some conditions.

Glassy phosphates. Glassy phosphates (polyphosphates) are useful in the chemical treatment of wells because of their ability to disperse iron oxides, manganese oxides, silts, and clays. Iron oxides are frequently responsible for the incrustation and clogging of the formation sand and the well screen. Treatment with a polyphosphate coupled with vigorous agitation is effective in removing this kind of incrustation. An important advantage of this treatment is that the chemicals are safe to handle.

The polyphosphates work in much the same way as household detergents, but do not form suds or foam. In well cleaning, foaming is not desirable because it interferes with mechanical agitation of the solution in the well. The polyphosphate solution does not dissolve the incrustation as acid does—fuming or boiling over does not occur. The polyphosphate cleans the well by breaking up the incrustation and dispersing it so that it can be readily pumped out of the well. Vigorous surging of the solution in the well is an essential part of the cleaning operation when polyphosphates are used.

A small amount of calcium hypochlorite, such as HTH or Perchloron, should always be used with the glassy phosphate. This chemical chlorinates the well and kills any iron bacteria or similar organisms that may be present.

Chlorine. In some locations where bacterial growths or slime clog the water-bearing formation, treating wells with chlorine has been found to be effective. The chlorine "burns up" the organic slimes responsible for the disruption in flow.

Chlorine must be used in concentrations strong enough to give a "shock" treatment. Concentrations of 100–150 mg/L of free chlorine are required. Calcium or sodium hypochlorite may be put into the well directly or in a water solution; chlorine gas is much more effective, but equipment for handling and proportioning it is not always available. Because chlorine gas is very corrosive and dangerously toxic when inhaled, it must be put into the well in a water solution.

Three or four successive chlorine treatments should be performed on an incrusted well. With repeated treatments there is a better chance that the chemical solution will be flushed through every part of the formation around the well that may be plugged. In some instances, alternate chlorine and acid treatments are highly effective.

Treating rock wells. The chemical cleaning procedures just discussed have more application to screened sand and gravel wells than to those drawing on solid-rock aquifers. Although proper acid treatment has proven to be a satisfactory maintenance procedure, in many rock formations the most satisfactory method of rehabilitating rock wells has been through the use of explosives.

Over the years, all sizes of charges have been used. However, collected samples and experience have shown that most plugging does not extend more than $1/4$ to $1/2$ in. (6 to 12 mm) back from the bore face. Therefore, the lightest charge that will effectively spall off that thickness of rock accomplishes an adequate cleaning; light charges

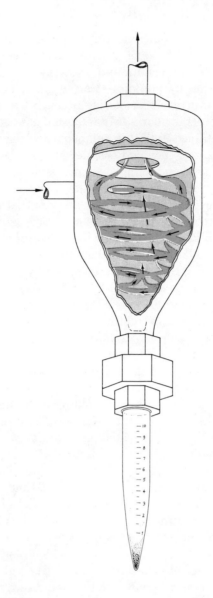

Figure 7-2 Centrifugal sand sampler.

also minimize the increase in bore size and reduce the charge sizes required for future cleaning.

Vibratory explosives. Vibratory explosives, used in new well development procedures, are worthy of consideration in maintenance work. This method is applicable to both screened and rock wells, but it is most effective when used as a first step in cleaning screened wells. Subsequent steps, such as acidizing, surging, and jetting, must be used to further break down and remove the particles produced by the explosion.

SAND PUMPING

As in the case of new well construction, it is important to clean a well thoroughly during maintenance or cleaning. Even small amounts of sand (as little as 0.3 ft^3/mil gal of water) can cause many operational problems. In addition to causing excessive wear in pumps and valves, control orifices can become plugged, water meters stopped, and sprinkler heads clogged.

Unfortunately, there are a few instances in which sand pumping cannot be eliminated, even if a well is properly designed and constructed. Where sand pumping cannot be eliminated at the design or desired capacity of pumping, a reduction of pumping rate, by increasing head through valving, may give some relief. Where water is discharged from the well into a large tank or reservoir, the sand may settle out and not cause excessive problems, except for pump and valve wear.

If sand pumping cannot be entirely eliminated, a centrifugal sand sampler, as shown in Figure 7-2, may be used. Water enters the body of the device at a tangent immediately below the baffle. The small radius and high velocity create a large centrifugal acceleration, which throws the sand to the side of the device. The sand falls down the side and is collected in the centrifuge tube, while the sand-free water flows out through the hole in the center of the baffle.

The flow is maintained at a constant rate, independent of the inlet pressure, by means of a flow-control valve that is rated at 0.5 gpm (2 L/min). The patented flow-control valve contains a rubber orifice that contracts with increasing inlet pressure in order to maintain a constant flow. According to the manufacturer, it is designed for a pressure variation from 15 to 150 psi (100 to 1000 kPa).

At suitable intervals, the volume of sand collected is recorded, together with the number of hours of operation. From these data, the average sand concentration may be computed, since the flow through the tester is known. Any significant increase in sand production is noted immediately, and corrective action can be taken before appreciable quantities of sand have entered the distribution system. The low cost of these testers makes it feasible to provide one for each well suspected of producing excessive sand.

AWWA MANUAL M21

Chapter 8

Groundwater Quality and Contamination

The chemical quality of groundwater is, for most uses, as important as the quantity of supply. Groundwater near the land surface, at depths of 500 ft (150 m) or less, ordinarily is subject to active replenishment and circulation. Shallow groundwater, therefore, commonly contains only small to moderate concentrations of mineral salts in solution.

In most locations, groundwater is also clean and biologically pure. The soil and rocks through which groundwater flows tend to remove bacteria and other organisms from the water by physical and biochemical processes. In addition, the depleted oxygen content and reducing conditions characteristic of the groundwater environment below near-surface levels are unfavorable to the survival of living organisms.

Unfortunately, today's groundwater supplies are being contaminated by man-made sources. These man-made contaminants will be discussed later in this chapter, but first a discussion of natural substances in groundwater will be presented.

NATURAL CHEMICALS IN GROUNDWATER

Precipitation, which travels overland to streams or infiltrates below the land surface to become soil water and groundwater, dissolves rock minerals and organic matter in the process. Consequently, the concentration of dissolved minerals in the water increases as the solution moves toward a balanced chemical state.

In some instances, the amount of dissolved minerals in the water can reach an excess. For example, brine, which usually occurs at great depths, contains so much dissolved mineral matter that the water is unsuited for virtually all purposes. The principal mineral constituents of natural brine, including mineral-laden thermal springs, are identified in Table 8-1. Groundwater brine has little or no worth, except as a source of certain minerals of commercial value. Because of its limited value, and its occurrence in deep geological formations that may be hydraulically insulated from

Table 8-1 The Principal Natural Chemical Constituents in Water, Concentrations, and Effects of Usability

Constituent	Concentrations in Natural Water	Effects on Usability of Water
Silica (SiO_2)	Ranges generally from 1.0 to 30 mg/L, although as much as 100 mg/L is fairly common; as much as 4000 mg/L is found in brines.	In the presence of calcium and magnesium, silica forms a scale in boilers and on steam turbines that retards heat and fluid flow; the scale is difficult to remove. Silica may be addded to soft water to inhibit corrosion of iron pipes.
Iron (Fe)	Groundwater having a pH less than 8.0 may contain 10 mg/L; rarely as much as 50 mg/L may occur. Acid water from thermal springs, mine wastes, and industrial wastes may contain more than 6000 mg/L.	More than 0.1 mg/L precipitates after exposure to air; causes turbidity, stains plumbing fixtures, laundry and cooking utensils, and imparts objectionable tastes and colors to foods and drinks. More than 0.2 mg/L is objectionable for most industrial uses.
Manganese (Mn)	Generally 0.20 mg/L or less. Groundwater and acid mine water may contain more than 10 mg/L. Water at the bottom of a stratified reservoir may contain more than 150 mg/L.	More than 0.2 mg/L precipitates on oxidation; causes undesirable tastes, deposits on foods during cooking, stains plumbing fixtures and laundry, and fosters growths in reservoirs, filters, and distribution systems. Most industrial users object to water containing more than 0.2 mg/L.
Calcium (Ca)	Averages about 15 mg/L in surface water, higher in groundwater. As much as 600 mg/L in some western streams; brines may contain as much as 75,000 mg/L.	Calcium and magnesium combine with bicarbonate, carbonate, sulfate, and silica to form heat-retarding, pipe-clogging scale in boilers and in other heat-exchange equipment. Calcium and magnesium combine with ions of fatty acid in soaps to form soap suds; the more calcium and magnesium, the more soap required to form suds. A high concentration of magnesium has a laxative effect, especially on new users of the supply.
Magnesium (Mg)	As much as several hundred milligrams per litre in some western streams; ocean water contains more than 1000 mg/L and brines may contain as much as 57,000 mg/L.	
Sodium (Na)	As much as 1000 mg/L in some western streams; about 10,000 mg/L in sea water; about 25,000 mg/L in brines.	More than 50 mg/L sodium and potassium in the presence of suspended matter causes foaming, which accelerates scale formation and corrosion in boilers. Sodium and potassium carbonate in recirculating cooling water can cause deterioration of wood in cooling towers. More than 65 mg/L of sodium can cause problems in ice manufacture.
Potassium (K)	Generally less than about 10 mg/L; as much as 100 mg/L in hot springs; as much as 25,000 mg/L in brines.	

Adapted from: Durfor, C.N. and Becker, Edith. Public Water Supplies of the 100 Largest Cities in the United States, 1962. USGS Water-Supply Paper 1812, Table 2, p. 16–19 (1964).

Table continues on next page.

Table 8-1 The Principal Natural Chemical Constituents in Water, Concentrations, and Effects of Usability (continued)

Constituent	Concentrations in Natural Water	Effects on Usability of Water
Carbonate (CO_3)	Commonly 0 mg/L in surface water; commonly less than 10 mg/L in groundwater. Water high in sodium may contain as much as 50 mg/L of carbonate.	Upon heating, bicarbonate is changed into steam, carbon dioxide, and carbonate. The carbonate combines with alkaline earths—principally calcium and magnesium—to form a crust-like scale of calcium and magnesium carbonate that retards flow of heat through pipe walls and restricts flow of fluids in pipes. Water containing large amounts of bicarbonate and alkalinity is undesirable in many industries.
Bicarbonate (HCO_3)	Commonly less than 500 mg/L; may exceed 1000 mg/L in water highly charged with carbon dioxide.	
Sulfate (SO_4)	Commonly less than 1000 mg/L except in streams and wells influenced by acid mine drainage. As much as 200,000 mg/L in brines.	Sulfate combines with calcium to form an adherent, heat-retarding scale. More than 250 mg/L is objectionable in water in some industries. Water containing about 500 mg/L of sulfate tastes bitter; water containing about 1000 mg/L may be cathartic.
Chloride (Cl)	Commonly less than 10 mg/L in humid regions; tidal streams contain increasing amounts of chloride (as much as 19,000 mg/L) as the bay or ocean is approached. About 19,300 mg/L in sea water; and as much as 200,000 mg/L in brines.	Chloride in excess of 150 mg/L imparts a salty taste. Concentrations greatly in excess of 150 mg/L may cause physiological damage. Food processing industries usually require less than 250 mg/L. Some industries—textile processing, paper manufacturing, and synthetic rubber manufacturing—desire less than 100 mg/L.
Fluoride (F)	Concentrations generally do not exceed 10 mg/L in groundwater or 1.0 mg/L in surface water. Concentrations may be as much as 1600 mg/L in brines.	Fluoride concentration between 0.6 and 1.7 mg/L in drinking water has a beneficial effect on the structure and resistance to decay of children's teeth. Fluoride in excess of 1.5 mg/L in some areas causes mottled enamel in children's teeth. Fluorides in excess of 6.0 mg/L causes pronounced mottling and disfiguration of teeth.
Nitrate (NO_3)	In surface water not subjected to pollution, concentration of nitrate may be as much as 5.0 mg/L but commonly is less than 1.0 mg/L. In groundwater the concentration of nitrate may be as much as 1000 mg/L where polluted, but generally less than 50 mg/L.	Water containing large amounts of nitrate (more than 100 mg/L) is bitter tasting and may cause physiological distress. Water from shallow wells containing more than 45 mg/L has been reported to cause methemoglobinemia in infants. Small amounts of nitrate help reduce cracking of high-pressure boiler steel.
Dissolved solids	The mineral constituents dissolved in water constitute the dissolved solids. Surface water commonly contains less than 3000 mg/L; streams draining salt beds in arid regions may contain in excess of 15,000 mg/L. Groundwater commonly contains less than 5000 mg/L and most of it at shallow depths contains less than 1000 mg/L; some brines contain as much as 300,000 mg/L.	More than 500 mg/L is undesirable for drinking and many industrial uses. Less than 300 mg/L is desirable for dyeing of textiles and the manufacture of plastics, pulp paper, rayon. Dissolved solids cause foaming in steam boilers; the maximum permissible content decreases with increases in operating pressure.

fresh groundwater resources, zones of brine in the subsurface are used for deep disposal of liquid wastes through injection wells.

Groundwater Use and Quality

The 14 principal chemical constituents of natural surface and groundwater are listed in Table 8-1. Only small to moderate amounts of these substances occur in most fresh water, as noted. Moderate amounts of dissolved minerals in water are desirable—water is more palatable (mineral-free water tastes flat to most people); minerals are important to human health, and plant and animal growth; and minerals reduce the corrosiveness of the water to pipelines and storage tanks. The effects of mineral constituents on water use are described in general terms in Table 8-1.

The quality of groundwater, including chemical, physical, biological, and radiological considerations, varies widely, and this variation is dependent on the type of water use. For example, the criteria for safe and healthy drinking water are much more stringent than for water used for industrial and agricultural purposes. Standard methods for conditioning groundwater to meet specified quality standards are described in chapter 10.

Chemical and Physical Characteristics

In addition to the effects of the chemical constituents listed in Table 8-1, groundwater possesses chemical and physical characteristics, which are imparted by certain chemical combinations of constituents, and by the physical and geochemical setting in which the groundwater occurs. The most common of these are hydrogen-ion concentration (pH), temperature, hardness, and gas content, which are discussed below.

Hydrogen-ion concentration (pH). A molecule of water (H_2O or HOH) in solution may partly dissociate into hydrogen (H^+) and hydroxyl (OH^-) ions. This form of ionization occurs only to a slight degree in pure water, and the H^+ concentration is only about 0.0000001 mol/L, or 10^{-7} mol/L. For convenience of expression, the negative exponent (or logarithm to the base 10) is used to express the hydrogen-ion concentration, or pH. Thus, the pH of the neutral solution (pure water) is expressed as pH 7. An H^+ concentration of 10^{-6} mol/L is expressed as pH 6, and it indicates that there are 10 times as many dissociated H^+ present as in the neutral solution with a pH of 7. On the other side of the neutral pH value, an H^+ concentration of 10^{-8} mol/L, or pH 8, indicates one tenth the number of dissociated H^+ present as in the pH 7 solution. Values of pH below 7 indicate acidity; those above 7 indicate surplus OH^- and basic solutions.

Most groundwater in the United States and throughout the world has pH values ranging from 6.0–8.5. Groundwater having a pH greater than 9.0 is unusual; free OH^- is rare in natural (uncontaminated) groundwater. Many thermal springs yield water with pH lower than 6, and river water unaffected by contaminants generally has a pH between 6.5 and 8.5.

Special techniques are necessary to accurately measure pH (Wood 1976). On-site measurement with carefully calibrated instruments and close adherence to prescribed methods yields reliable results. Determination of the pH of water pumped from a well requires sampling precautions to ensure that the condition of samples brought to the surface for measurement are representative of the physical (temperature and pressure) and chemical states of water residing in the aquifer.

Temperature. In contrast to the seasonal and diurnal fluctuations of surface-water temperature, the temperature of groundwater is virtually constant. The exception is the temperature of groundwater near the land surface, which may fluctuate

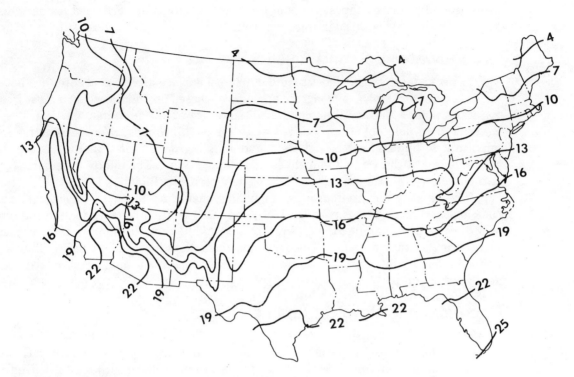

From Basic Ground-Water Hydrology. R.C. Heath. US Geol. Surv. Water-Supply Paper 2220, US Govt. Printing Ofce. (1983).

Figure 8-1 Approximate temperature of groundwater (in degrees Celsius) in the continental United States at depths of 30 to 80 ft.

several degrees during the year in response to seasonal insolation effects. The constancy of groundwater temperature is important to the palatability of drinking water.

The approximate temperature of groundwater 30–80 ft (9–24 m) below the land surface in the United States is shown by temperature contours in Figure 8-1. The mean temperature at these shallow depths generally is on the order of 2 to 3°F (−17 to −16°C) above the mean annual air temperature. Below this zone of solar influence, the temperature of groundwater increases at a rate of approximately 1°F (0.6°C) for each 64 ft (20 m) of depth. This rate is in accordance with the geothermal gradient of the earth's crust. Thus, with few exceptions, groundwater pumped from deep wells is of higher temperature than that pumped from shallow wells.

Hydrogeological settings that accelerate or retard groundwater flow may disrupt the uniform temperature gradient with depth. Deep, rapid circulation of relatively cool, shallow groundwater, for instance, may modify the normal gradient. Waste-heat discharge to the ground can create local zones of abnormally high groundwater temperature.

The characteristics of groundwater temperature make groundwater systems a valuable resource today. Geothermal groundwater systems have attracted considerable interest during the past several decades as potential sources of heat and power for municipal and industrial usage in the United States. Geothermal energy is captured by removal of heat from deep-lying groundwater pumped to the surface. Some heat-pump installations are designed to take advantage of the relatively small temperature difference between shallow groundwater and the atmosphere, providing economical heating and cooling systems for domestic and small commercial and industrial requirements.

Hardness. Hardness is a property of water derived mainly from the presence of calcium and magnesium, although other divalent metallic cations may also contribute hardness to the water. In water, these metallic ions inhibit lathering, react with soap to form undesirable precipitates, and combine with certain anions in boiler water to form efficiency-robbing scale on tank walls and in pipes.

In aquifers containing hard water, lowering the water level in a well during pumping and the corresponding reduction in water pressure at the intake screen may precipitate calcium and magnesium compounds that clog the well screen. Similarly, in an unscreened well the precipitates may clog the openings in the aquifer immediately adjacent to the well bore with comparable reduction in inflow of water.

A number of similar numerical scales for rating water hardness have been devised and published, including, for example, the following scale (Durfor and Becker 1964):

Hardness Range mg/L of $CaCO_3$	Description
0–60	Soft
61–120	Moderately hard
121–180	Hard
More than 180	Very hard

Hardness of water used for domestic purposes is not objectionable in concentrations less than about 100 mg/L. The hardness of groundwater throughout much of the United States is less than 100 mg/L. However, groundwater in gypsiferous and carbonate bedrock formations of the north central region (including North Dakota, South Dakota, Iowa, and parts of surrounding states) and groundwater in other parts of the nation that is underlain by sedimentary rocks rich in calcium and magnesium generally exceed this level. In these areas, hardness levels of 300 mg/L are common, and levels as high as 1000 mg/L occur in some places. Where hardness is a serious problem, effective water-softening equipment is available, although it is costly to operate for treatment of large volumes of water.

Gases

Precipitation is the source of most groundwater. As precipitation falls, it comes in contact with soluble gases that may combine with the water droplets; it also contacts dust and other particulate matter suspended in the air that add chemical constituents to the water. Combination of precipitation with carbon dioxide (CO_2) to form carbonic acid (H_2CO_3) is the most significant atmospheric influence on the acidity of precipitation. Additional carbon dioxide, originating from organic processes at the land surface and in the soil zone, dissolves in groundwater, further increasing its acidity and its capacity to dissolve mineral matter.

Oxygen (O_2) and hydrogen sulfide (H_2S) are other important gases occurring in groundwater. Dissolved oxygen should be distinguished from oxygen that is chemically bonded to hydrogen to form water. The concentration of dissolved oxygen in shallow groundwater is usually less than 10 mg/L, and in deep-lying groundwater it may be virtually absent. It is harmless to health and its presence may improve the palatability of water. However, dissolved oxygen does contribute to water's corrosiveness to metals, most aggressively where carbon dioxide or low pH or both are present. Hydrogen sulfide gas is generated in groundwater by the decomposition of natural organic substances and by the action of sulfate-reducing bacteria on organic materials. Hydrogen sulfide is corrosive in the gaseous state and in combination with water, with which it forms a weak acid solution.

Methane (CH_4), generated by the decomposition of vegetation and other organic materials, is prevalent in groundwater and soil moisture zones in small concentrations. Larger quantities, sufficient for domestic or small industrial heating and energy requirements, may be formed in peat bogs, coal mines, or large landfills containing thick deposits of decomposing organic wastes. Fires and explosions in mines, basements, water wells, and petroleum wells often are attributable to the accumulation of methane gas, released from groundwater reservoirs or rock and coal formations.

GROUNDWATER CONTAMINATION

Contamination of groundwater is a widespread and challenging problem in the United States and many other countries. Unfortunately, contaminants flow undetected into groundwater reservoirs, sweeping through the aquifer until a sizable portion of the aquifer has become degraded, often beyond practical rehabilitation. Even where physically and economically practical, rehabilitation of a contaminated aquifer is a difficult, complicated, and expensive undertaking.

Realization of often irreversible damages to the nation's groundwater resources has stimulated public efforts to reduce the influx of contaminants at their sources. These actions are based on the conviction that prevention is simpler, more effective, and less costly than clean-up measures. It is important to recognize, however, that even after elimination of contamination sources, aquifers may remain contaminated for centuries due to the slow rate of groundwater movement and the corresponding slow rate of dilution and flushing of contaminating substances.

Sources of Contaminants

Potential contaminants are generated by virtually all of man's industrial, agricultural, urban, and rural activities. Therefore, it is unrealistic to assume that all groundwater contamination could be prevented completely. The principal sources of contaminating substances in the United States, listed in order of national importance, are given below (Source: American Institute of Professional Geologists 1983).

- industrial waste;
- municipal landfills;
- agricultural chemicals;
- septic-tank and cesspool effluents;
- leaks from petroleum pipelines and storage tanks;
- animal wastes (feedlots);
- acid mine drainage;
- oil-field brine;
- saline water intrusion; and
- irrigation return flow.

For a specific region or locality, the order will be somewhat different. It is obvious from the list above that a wide variety of contaminants find their way to groundwater reservoirs. These include chemical, physical, biological, and radiological contaminants, which are generated most prolifically in industrialized, economically advanced nations such as the United States. Table 8-2 lists the maximum contaminant levels established for drinking water in the United States. The table reflects

Table 8-2 Maximum Contaminant Levels for a Variety of Organic and Inorganic Chemicals

Constituent	Maximum Concentration (in mg/L unless specified)
Arsenic	0.05
Barium	1.
Cadmium	0.010
Chromium	0.05
Lead	0.05
Mercury	0.002
Nitrate (as N)	10
Selenium	0.01
Silver	0.05
Fluoride	4.0
Total THMs	0.1
Turbidity (surface-water systems only)	1 NTU to 5 NTU
Coliform bacteria	1/100 mL (mean)
Endrin	0.0002
Lindane	0.004
Methoxychlor	0.1
Toxaphene	0.005
2,4-D	0.1
2,4,5-TP (Silvex)	0.01
Combined Radium-226 and -228	5 pCi/L*
Gross Alpha	15 pCi/L
Beta particle and Photon activity	4 millirem/yr†
Trichloroethylene	5 µg/L
Carbon tetrachloride	5 µg/L
Vinyl chloride	2 µg/L
1,2-Dichloroethane	5 µg/L
Benzene	5 µg/L
p-Dichlorobenzene	75 µg/L
1,1-Dichloroethylene	7 µg/L
1,1,1-Trichloroethane	200 µg/L

*Picocuries per litre.
†Annual dose equivalent to the body or any internal organ.

the wide spectrum of organic and inorganic chemicals considered to be toxic to humans and requiring surveillance and regulatory measures.

Contaminants may enter groundwater reservoirs by intent and design, such as deliberate placement in the subsurface through a waste-injection well. Or, contaminants may enter a groundwater system inadvertently, for example, by leakage from a ruptured pipeline, as leachate from an inadequately sealed landfill, or as a product of agricultural fertilizer application.

Figure 8-2 illustrates the downward movement of a contaminant from a land surface source (in this example, a waste lagoon), through the zone of aeration (the zone of rock and soil above the water table, it is unsaturated with water), and then into the aquifer. As illustrated, the contaminating liquid migrates downward and laterally into the aquifer. The recharge mound built up beneath the source of the contaminant may propel some of the liquid in an upward direction (to the left in the illustration) for relatively short distances, but the dominant direction of movement for contaminant liquids with density similar to that of the groundwater is downward in the direction of general groundwater flow, indicated by the direction of slope of the old water table (to the right in the illustration).

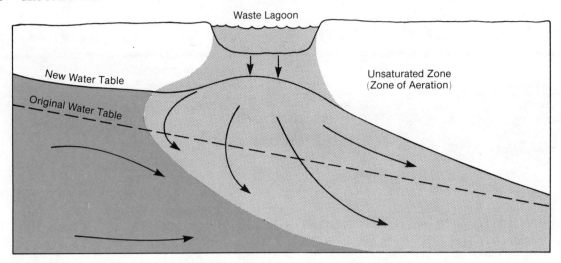

Figure 8-2 Flow of contaminants from a ponded surface source into an aquifer.

The relative densities of incoming fluids and of the receiving water in the aquifer influence the pattern of contaminant movement. Low-density fluids, such as gasoline and oil, tend to "float" on the groundwater and perhaps migrate upward. Comparatively, high-density fluids, such as brines, tend to "sink" to the lower part of the aquifer and migrate downward.

MANAGEMENT OF GROUNDWATER QUALITY

Until the 1960s, water-resources management in the United States focused primarily on the demands for quantities of water, with only small interest or effort being devoted to water quality. In subsequent years, as water-quality problems became more serious, growing attention has been devoted to water-quality concerns, first to the improvement of surface-water quality and, only more recently, to groundwater quality. Technical methods for evaluating and managing groundwater flow and supply are well advanced, and these methods are summarized in chapters 2 and 4. Parallel methods are now evolving for managing and protecting the groundwater quality, including conjunctive management of both quantity and quality of the resource. Many textbooks and governmental publications describe the techniques available to investigate, plan, manage, and regulate groundwater quality (Freeze and Cherry 1979; Hem 1970).

References

DURFOR, C.N. & BECKER, EDITH. Public Water Supplies of the 100 Largest Cities in the United States. US Geol. Surv. Water-Supply Paper 1812, US Govt. Printing Ofce. (1964).

FREEZE, R.A. & CHERRY, J.A. Groundwater. Prentice-Hall Inc., New York (1979).

Ground Water—Issues and Answers. Amer. Inst. Prof. Geol., Arvada, Colo. (1983).

HEM, J.D. Study and Interpretation of the Chemical Characteristics of Natural Water. US Geol. Surv. Water-Supply Paper 1473, US Govt. Printing Ofce. (1970).

WOOD, W.W. Guidelines for Collection and Field Analysis of Ground-Water Samples for Selected Unstable Constituents. US Geol. Surv. Tech. of Water Res. Invest. Book 1, Chap. D2 (1976).

AWWA MANUAL M21

Chapter 9

Groundwater Treatment

Groundwater is usually well-filtered and free from turbidity, color, and organic chemicals, as a result of extensive slow percolation through the earth. Unless the mineral or dissolved gas content is excessive, well water is usually considered quite palatable and often is preferred over surface water for drinking. Excessive mineral content of groundwater (discussed in chapter 8), however, frequently makes conditioning desirable for most domestic and industrial purposes.

The most widely applied methods of treating groundwater will be discussed in this chapter. Methods that apply primarily to surface or saline waters may be mentioned for comparison. Methods in use for very specific problems will not be covered.

AERATION

Aeration is the exposure of water to air for the purpose of removing undesirable gases, such as free carbon dioxide, hydrogen sulfide, volatile organic chemicals, and methane. Aeration is also used to remove iron and/or manganese from well-water supplies. These two elements are usually found in well water in the form of ferrous (Fe^{++}) and manganous (Mn^{++}) bicarbonate, because most well waters lack dissolved oxygen.

The presence of iron and/or manganese in the unoxidized state is noticed when freshly pumped well water turns milky and yellowish to blackish on exposure to air. This is due to the oxidation of the iron and manganese and the resulting formation of the insoluble ferric hydroxide and manganese dioxide. Therefore, aeration must be followed by sedimentation and filtration to remove the precipitate.

Treatment Methods

The most common aeration methods are

- introduction of air into water, and
- introduction of water into air.

In the first method, also called air diffusion, compressed air is injected into an open or closed tank, which may or may not be vented. The efficiency of this method is determined by several factors, such as depth and cross-sectional area of tank, type of air diffusers, and air flow.

In the second method, water is either sprayed or uniformly distributed above a series of staggered slat trays, trays of coke, or some manner for allowing water to cascade down the aerator. The water is broken up into small droplets or thin films and exposed to a controlled countercurrent flow of air by natural or forced draft.

Although aeration is by far the most common method of oxidizing iron and manganese for removal from groundwater supplies, strong oxidants, such as chlorine, chlorine dioxide, potassium permanganate, and ozone, can be used effectively in the removal of these two troublesome elements. Sodium cation exchangers also remove dissolved ferrous and manganous ions by ion exchange, simultaneously with hardness.

SOFTENING

Most groundwater contains significant quantities of dissolved calcium and magnesium salts in the form of bicarbonates, sulfates, chlorides, and nitrates. It is the presence of these ionized compounds that accounts for most hardness in water. As mentioned earlier, hardness causes curdling of soap and resultant increases in its consumption; deposits of scale in plumbing, boilers, and water heaters; problems in some industrial processes, such as bottle washing, food processing, sterilization, pasteurization, photography, and pulp and paper production; and unpleasant tastes of many foods.

There are two forms of hardness—carbonate and noncarbonate. Waters containing either calcium or magnesium bicarbonate are said to have carbonate or temporary hardness, since the hardness may be removed to the limit of solubility simply by boiling the water. Noncarbonate or permanent hardness refers to the calcium and magnesium associated with sulfates, chlorides, and nitrates.

Treatment Methods

Softening of hard water is usually accomplished by the following two methods:
- chemical precipitation with lime alone or lime and soda ash (lime–soda ash process), and
- sodium-cycle ion-exchange process.

Lime–soda ash process. The lime–soda ash process can be performed at ambient temperatures, in which case it is called "cold process," or at elevated temperatures, in which case it is called "hot process." Cold lime softening is one of the oldest methods of water treatment. It consists of adding lime (calcium hydroxide) or lime in combination with soda ash (sodium carbonate) to reduce water hardness. The hot process is used primarily by industry for the treatment of medium-pressure boiler feedwater. The choice of treatment depends on the composition of the water to be treated and the degree of hardness reduction desired.

Effluent. Temperature, retention time, and contact of previously formed precipitates with influent raw water and treatment chemicals will influence chemical efficiency and the effluent water quality of the cold lime–soda ash softening process. A properly operated cold softener has the capability to produce the following effluent: calcium—35 mg/L as $CaCO_3$; total alkalinity—35 mg/L as $CaCO_3$; and carbon dioxide—0.

The effluent from a lime or lime–soda ash process softener (either hot or cold) is saturated or supersaturated with calcium carbonate, making this effluent scale-forming. Consequently, post-precipitation of calcium carbonate may occur in the filters and in the piping system. Stabilization of the effluent with carbon dioxide (recarbonation) or acid treatment is required to convert carbonates to bicarbonates and to render the water stable. Magnesium will be reduced to 50 mg/L as $CaCO_3$ if present in the raw water in an amount greater than 50 mg/L or reduced by 10 percent if present in lesser amounts.

Often calcium carbonate and magnesium hydroxide precipitate in the form of very fine particles, almost colloidal in size, which will not settle in any reasonable time. Therefore, coagulants such as ferrous or ferric sulfate, alum, and polyelectrolytes, must be added to agglomerate these particles into large aggregates that will settle more easily.

Ion–exchange process. Ion exchange, as the name implies, is a chemical process that reversibly exchanges undesirable ions for ions that are not undesirable. Softening by the sodium-cycle ion-exchange process involves the exchanging of hardness ions, calcium and magnesium, for very soluble sodium ions.

In the early days of ion exchange, this softening process was called sodium zeolite softening, because natural or synthetic zeolites were used as ion exchange materials. These materials included material made by fusing kaolin clay, sand, and sodium carbonate; siliceous inorganic exchange material produced from the reaction of sodium silicate, sodium aluminate, and aluminum sulfate; natural inorganic siliceous greensand (glauconite); and sulfonated coal.

Today, most ion-exchange materials consist of a matrix or hydrocarbon network such as polystyrene, which is copolymerized with divinyl benzene (DVB). The matrix is converted to a strong-acid cation exchanger or to a strong-base anion exchanger, depending on the type of ionizable groups attached to the network. In softening by ion exchange, only strong-acid cation-exchange resins operating in the sodium cycle are used as exchange material.

The ion-exchange softening process consists of passing the hard water, usually under pressure, through a column (ion exchanger) containing the cation-exchange resin in the sodium form. In the column, calcium and magnesium ions are replaced with the more soluble sodium ions.

Other dissolved ions, such as iron, manganese, barium, strontium, and zinc, are also removed from solution and replaced with sodium ions during the softening process. These ions are normally present in such small amounts that they are usually neglected in hardness calculations.

FILTRATION

Filtration is the process used to remove suspended matter from water. In groundwater treatment, filtration is used to remove suspended solids that resulted from some previous treatment process, such as aeration, chemical oxidation of iron/manganese, or lime softening. Filtration essentially consists of passing the water (usually vertically downward) through some type of granular medium that removes the suspended solids.

Gravity and Pressure Filters

Basically, there are two types of filters—gravity and pressure. Gravity filters, constructed of steel or concrete and round or rectangular in shape, are open at the top and operate with only the pressure provided by the height of the water above the filter media. For large installations, rectangular concrete filters are preferred because they may be built with common walls between them.

The filter bed in a gravity filter is similar to that in a pressure filter. It consists of graded filter sand, anthracite coal, or a combination of the two. The filter media is usually supported by graded gravel, which is underlain by an underdrain system that collects the filtered water and distributes the backwash water.

Pressure filters are of closed construction, having a vertical or horizontal cylinder of iron or steel. The filter operates under the pressure of the influent water. Pressure filters are preferred for small flows and industrial applications. These filters have the advantage that they can be operated at higher pressure drops and need no repumping.

Backwashing

During the filtration process, the suspended matter accumulates in the filter media, creating a resistance to flow over and above the normal resistance of the clean bed. As filtration proceeds, the pressure drop across the bed continues to build up until it reaches a preset point, which determines the end of the filtering cycle. The filter is then backwashed by allowing the backwash water to flow back through the underdrain system. The backwash rate should be high enough to expand the filtering media and dislodge the entrapped solids.

GRANULAR ACTIVATED CARBON TREATMENT

Granular activated carbon (GAC) is used as an adsorbent when treating groundwater. It is effective in removing or reducing concentrations of some organics and total organic carbon. Granular activated carbon is able to remove organics because of the large surface area of the carbon. Water is passed through the carbon beds and the organics become affixed to the carbon. The process is expensive because of the cost of replacing and regenerating the carbon; the carbon must be replaced periodically and/or reactivated by thermal treatment to drive off the organics.

CHLORINATION

Chlorination is the process of applying chlorine to water or to waterborne wastes for the purpose of disinfection or to accomplish certain biological or chemical changes. Groundwater supplies are chlorinated to destroy harmful organisms, to ensure disinfection, to control or remove tastes and odors, and to oxidize hydrogen sulfide, iron, and manganese.

Chlorine Residuals

One of the outstanding features of chlorination, when compared to other methods of disinfection, is that its effectiveness can be determined. When the presence of residual chlorine in water can be demonstrated, the effectiveness of chlorination is definitely ensured, and continues as long as the residual is present.

The different chlorine compounds that may be found as residual chlorine in water include hypochlorous acid (HOCl), hypochlorite ion (OCl^-), and chloramines. Residuals consisting of hypochlorous acid and hypochlorite ions are defined as free available chlorine. Chloramines or a chemical combination of chlorine with ammonia or organic nitrogen are defined as combined available chlorine.

Chlorine Compounds and Equipment

The size of the system, raw-water quality, extent of conditioning needed, and availability usually dictate the type of chlorine compound to be used. Where available and where use warrants, elemental chlorine is commonly used, but calcium (solid form) or sodium (solution form) hypochlorite are satisfactory for use in small systems.

Elemental chlorine is a liquefied gas under pressure. It is generally available in 100- and 150-lb (45- and 70-kg) cylinders and in 2000-lb (900-kg) containers. The chlorine in containers has both a liquid and a gas phase. Single-unit tank cars are also available, with capacity of 16, 30, 55, 85, and 90 tons (15,000; 27,000; 50,000; 77,000; and 82,000 kg).

Chlorine gas is fed with a solution-feed chlorinator that operates under vacuum to minimize the possibility of gas leaks. The vacuum is developed by an aspirator-type injector operated by water under pressure, the chlorine gas is regulated and metered and dissolved in water at the injector. This chlorine solution is then discharged to the point of application.

Calcium hypochlorite is a dry, granular material available in powdered form or in compressed tablets. Sodium hypochlorite is usually supplied as a transparent, light-yellow solution of sodium hypochlorite and water. The strength, or oxidizing power, of these two compounds is expressed in terms of available chlorine. Calcium hypochlorite and sodium hypochlorite contain approximately 70 and 15 percent available chlorine, respectively. Household bleach, a form of sodium hypochlorite, contains 5 percent available chlorine. Hypochlorites are generally fed as water solutions using positive-displacement metering pumps or any other controlled metering device.

Although chlorine or chlorine compounds are the most commonly used disinfectants, other methods of disinfecting water have been developed. These include ozonation, ultraviolet radiation, and chlorine dioxide application.

FLUORIDATION

Fluoride in drinking water in concentrations of approximately 1.0 mg/L offers protection against dental cavities and will enhance the formation of dental enamel. However, water supplies with excessively high natural fluoride levels have been shown to be associated with dental malformation, stained enamel, and other dental disorders. Because of this, the National Primary Drinking Water Regulations have established an optimum fluoride level for community water systems of 4.0 mg/L. Water supplies that exceed the maximum permissible levels for fluoride must reduce the concentration through defluoridation or by blending with other water sources that have a low fluoride content. Since groundwater generally contains only small concentrations of fluoride, fluoride is commonly added to groundwater supplies that will be used for potable service, to provide a residual of approximately 1.0 mg/L.

Chemicals and Equipment

The following chemicals are used in the process of fluoridation or artificial adjustment of fluoride levels in domestic water supplies: sodium fluoride, hydrofluosilicic acid, sodium silicofluoride, calcium fluoride, and ammonium fluosilicate. Standard chemical feeders, such as positive-displacement metering pumps and dry feeders (either gravimetric or volumetric), are generally used for feeding and metering fluoride compounds. The type of feeder selected depends on the fluoride compound used and the quantity to be fed.

AWWA MANUAL M21

Chapter **10**

Record Keeping

Proper operation of a groundwater system includes the gathering, compiling, and recording of a wide variety of data. Such records are invaluable in determining the time for required well and pump maintenance (before complete failures occur), costs of water production, proper amortization of investments, adequacy of source, and many other related and seemingly unrelated system factors.

RECORD-KEEPING OBJECTIVES

The basic objective of any record-keeping program is to compile information that makes it possible to compare actual operating characteristics with original and calculated design performances.

Information Needed

The data collected and compiled should include detailed individual well logs; pump-design particulars and maintenance records; water-quality analyses; individual well meter readings; pumping and nonpumping water levels; pump discharge pressures; changes in piping or other factors affecting head conditions; power meter readings; and as much additional data as may be available (Figures 10-1 and 10-2).

A newly constructed well with a new pump installed should be evaluated as a unit to provide a standard by which the performance at any future time is measured. After well and pump evaluation, data should be collected to determine well losses that relate to well design and construction. Comparative data pertaining to the physical condition of the pump unit should also be collected. Proper study and comparison of such data enable the operator to anticipate maintenance and repair needs. Of course, after any repair or maintenance work, similar tests should be rerun.

If original records have been lost or not kept at all, manufacturers and well drillers, who maintain itemized records, including details for original pumps sold and installed, can be contacted. Copies of those records give indications of the original pumping levels and head conditions. They may also indicate general area changes in water levels. Actual drilling logs may be replaced by running gamma-ray logs in both new and old wells and then compared to correlate formation compositions. Changes in water quality may be documented in state health agency files and be relevant to well maintenance.

WELL PUMPING RECORD

STATION ______ PUMP SETTING ______

STARTING PRESSURE ______ NONPUMPING LEVEL ______

DATE	TIME	AM / PM	ON / OFF	FLOW TOTALIZER	UNIT	ELAPSED TIME INDICATOR	TOTAL HOURS RUN	TOTAL WATER PUMPED	PUMPING LEVEL ft.	SHAFT OIL RATE dpm	PRESSURE psi	OPERATOR
				Net		Net						
				Net		Net						

DATE	TIME	AM / PM	POWER METER	FLOW RATE	REMARKS	INITIAL
			Net			
			x K =			
			K.W.H.			

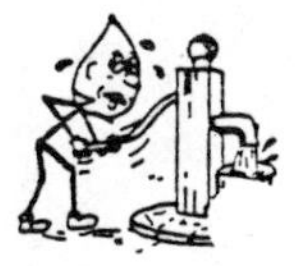

Supervisor

Figure 10-1 Sample well-pumping record form.

WELL PUMPING EFFICIENCY

DATE: ______

OPERATOR: ______

WELL	SAME PERIOD OF TIME: WATER PUMPING (MILLIONS GALLONS)	SAME PERIOD OF TIME: K. W. H.	STATIC LEVEL (FT.) IF POSSIBLE	PUMPING LEVEL (FT.)	PUMPING PRESSURE P. S. I.	TOTAL HEAD (FT.)	EFFICIENCY (%)
EXAMPLE	5.196	7370.6	162	181	44	282.64	62.6

$$\text{EFFICIENCY} = \frac{\text{M G} \times \text{TOTAL HEAD}}{.3182 \times \text{K. W. H.}}$$

TOTAL HEAD = (PUMPING LEVEL + (PUMPING PRESSURE X 2.31))

Figure 10-2 Sample well-pumping efficiency calculation form.

AWWA MANUAL M21

Chapter 11

Overview—Design, Construction, and Testing of Wells

In previous chapters, some design considerations regarding the physical properties of an aquifer were discussed. However, there are many additional factors that enter into judging the location, design, and construction of a well or well fields. These factors include, but are not limited to, sources of recharge and contamination, land use and development in the vicinity, operation and maintenance of the well units, construction economics of attendant facilities (pipe lines, power lines, and enclosures), and interference with other wells or well fields. The combination of factors affecting a particular area usually does not match that of another. Therefore, general discussion and guidelines are given that apply in some degree to all utilities using groundwater as their water source.

AQUIFER CHARACTERISTICS

For potable water to occur underground, there must be some recharge and discharge source in recent geologic time. Most fresh waters in aquifers vary from a few years old to a thousand or more years old. Water with high total dissolved solids content and water that is brackish is usually older or more removed from the recharge source. Generally, shallow groundwater is younger with less total dissolved solids than deeper groundwater. However, there exists deeper, permeable aquifers, which may underlie brackish water at a shallow depth, that are connected to a recharge and discharge source.

Recharge and Internal Storage

All aquifers have the capacity for recharge and internal storage. The simple analogy that best shows these two properties is to visualize a 1000-gal horse tank being fed by a 10-gpm flow from a garden hose. The tank is full and overflowing. If a 100-gpm water demand was taken from the tank, the tank could support this demand for approximately 10 min. This would imply a 100-gpm yield was available. If the pumping continued until the tank was emptied, it would be found that only 10 gpm, or the recharge rate to the tank, could be pumped.

Now let's apply this analogy to a groundwater aquifer. When an aquifer is full and overflowing, numerous springs and seeps may occur in areas of depression or low elevation, which allow the groundwater to come to the surface where it either evaporates or flows off as a stream or water course. Developing and pumping a large-capacity well field may depress the water table until the hydraulic gradient from a recharge source is in equilibrium with the average rate of withdrawal. If pumping exceeds the ability of the aquifer to transmit water from a recharge source to the discharge area, then dewatering of the aquifer will occur. If water levels in a well field are progressively lower year after year, the aquifer is being pumped in excess of recharge capacity and water is being removed from the aquifer's internal storage capacity. The sustained yield of an aquifer is the yield that can be withdrawn indefinitely without dewatering. Yields in excess of the sustained yield are derived from the storage capacity of the aquifer.

Storage and recharge rates. Aquifer storage and recharge rates vary greatly with the type of material in which the aquifer occurs.

Sands and gravels. Sands and gravels have a porosity of 30 to 35 percent of the total volume of saturated material. Three feet (1 m) of saturated sand and gravel can contain the equivalent volume of 1 ft (0.3 m) of water over the area. However, not all of this water will flow by gravity to wells. Molecular attraction and surface tension usually limit the effective yield to approximately 50 percent or less. Thus, 5–6 ft (1.5–2 m) of saturated sand will yield the equivalent of 1 ft (0.3 m) of water over the same surface area.

Often, unconsolidated aquifers occur near lakes, rivers, or streams. In these situations, wells taking water from the aquifers may affect the storage capacity only slightly because the wells derive their water supply through infiltration from the surface water source. Thus, unconsolidated sand and gravel aquifers have the ability to transmit large quantities of water from recharge sources.

Sand and gravel aquifers can easily supply a water yield to wells in excess of the recharge capacity from the aquifer's large internal storage capacity. Thus, peak water demands are easily met from this aquifer type.

Consolidated rocks. Consolidated rock aquifers consisting of limestones and dolomite often have a low porosity, usually 1 to 3 percent, and can only transmit water from an available recharge source to the withdrawal point or well. Thus, it could take from 30–100 ft (9–30 m) of saturated rock formation to hold the equivalent of 1 ft (0.3 m) of water over the surface area. These aquifer types can seldom be developed in excess of their transmission capacity or recharge rate. The exception is limestones in coastal plain areas, which are made up of small shells and shell fragments, identified as a reef. These formations exhibit characteristics somewhat similar to an unconsolidated aquifer.

Water Movement

It is commonly assumed that one obtains water from a well, but nothing could be further from the truth. A well is a means of access to a water-bearing formation, and it serves the same purpose as a straw in conducting fluid from a glass to your mouth. A well typically includes a pump, which moves water from the aquifer to a distribution system for delivery to the water user.

Cone of depression. To move water from a formation into a well, a gravitational force must be created. The gallonage first pumped from a well removes water in storage from the well bore, then removes water from storage in the aquifer. This creates a pressure sink that starts water movement through the formation material into the well bore. This pressure sink is commonly referred to as the cone of depression. If the aquifer is unconfined, as shown in Figure 11-1, then the water table surface within the grains of sand actually forms an inverted cone.

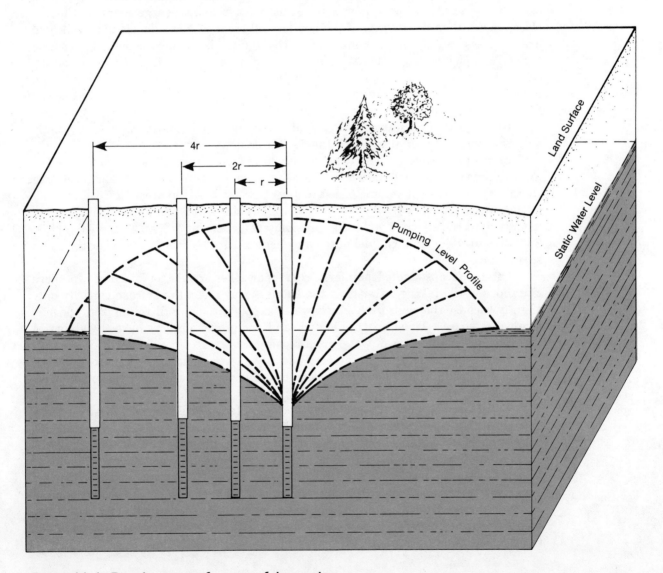

Figure 11-1 Development of a cone of depression.

The cone of depression is circular only if the formation material is homogeneous and isotropic, and the water level is level. As the slope of the water level becomes significant, the cone of depression around the well becomes egg shaped with the short axis up-gradient and the long axis down-gradient. Similarly, if a well was located near a riverbank with connection between the aquifer and the surface water, the cone of depression may extend only a short distance toward the river and extend a greater distance away from the recharge source.

Change in head. A basic assumption made in the theoretical development of groundwater evaluation formulas is that water is instantaneously released from storage with a change in head. Actually, this seldom occurs. Coarse-grained aquifers approach this theoretical assumption in that water is freer to flow through the pore space towards the wells when the pressure sink is created. As the aquifer becomes more fine grained, the movement from the aquifer to the well bore is slower because of the pressure sink. The cone of depression may develop rapidly to $1/2$ mi (1 km) or more as the water begins draining and migrating slowly to the well bore.

Hydraulic Conductivity and Transmissivity

Ideally, when developing a groundwater source all attendant facilities would be placed at one location and one large-capacity well would serve the needs of the utility system. However, this is seldom practical. Often, the cost of attendant facilities far exceeds the cost of well construction and receives extensive consideration when the number and capacity of wells are being selected.

Depending on the hydraulic conductivity and transmissivity of the aquifer, numerous wells may need to be drilled in order to meet the water supply needs. As wells are added within the radius of influence of each other, there is an overlapping or interfering drawdown profile (Figure 11-2). If wells are grouped closely to each other, they take on the characteristic of a large-radius singular well. The group of wells is then dependent on the storage and transmissive characteristics of the aquifer system.

WELL FIELD EVALUATION

When developing a groundwater supply in an undeveloped area, an extensive study needs to be made of the well field. This includes the aquifer material characteristics, the aquifer's recharge source or sustainable yield, the aquifer's natural discharge (if appropriate), the water quality characteristics, and current use being made of the water supply system by others.

Information Sources

Most major aquifers have now been mapped by the US Geological Survey and/or geologic and water resources agencies of state governments. Geologic and water supply reports generally exist for most regions of the United States and for many areas in foreign countries. A review of available literature and data is the first step in identifying the potential location and yield characteristics of a new well field.

After a potential area has been identified, permission should be obtained from appropriate landowners to conduct detailed geophysical surveys (if necessary), in order to identify the areal extent or location of the best aquifer for water yield. This is followed by test-hole drilling. Often, only small land parcels are acquired for the individual well sites. Appropriate spacing between land parcels makes it possible to obtain the required yield from a multiwell development of a large aquifer system.

126   GROUNDWATER

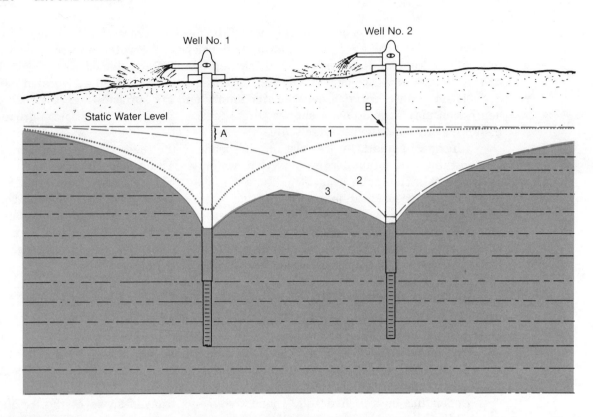

1—drawdown profile when well 1 is pumped alone; 2—drawdown profile when well 2 is pumped alone; 3—drawdown profile resulting from interference when both wells are pumped simultaneously; A—vertical distance that water level in well 1 drops as a result of pumping well 2; B—drop in water level in well 2 as a result of pumping well 1.

Figure 11-2 Section showing overlapping or interfering drawdown profiles.

For a new development, the investigation should extend beyond the immediate area contemplated for the well field. Test holes with monitoring wells should be installed both up-gradient and down-gradient and to either side of the site to verify the geologic conditions, verify and periodically monitor water quality, and monitor the head and drawdown characteristics, which later may be imposed in the vicinity because of well field development. Thus, a minimum of 8 to 10 test holes to more than 25 or 30 test holes may be drilled in a given area to determine the areal extent of the geologic formation, hydraulic gradient, and water quality surrounding the proposed area.

Test Wells

The investigation's next phase may warrant construction of a temporary test well with an appropriate piezometer array installed for shallow wells. The piezometer array will be used to collect data as tests are carried out on the test well. They are used to measure the in situ hydraulic conductivity, formation transmissivity, storage coefficient, and potential sustainable yield of an individual well in the aquifer or aquifers selected. In some cases, the actual aquifer test conducted with the test well may extend from one day to a month or more to fully assess the yield characteristics before a major investment is made in the area. Water quality parameters may be sampled frequently during the test pumping period to see if any significant changes in water quality occur as a result of the new pumping stress being imposed on the aquifer.

Water levels are monitored in observation wells in the immediate drawdown influence of the test well to determine the aquifer flow characteristics. At the conclusion of actual pumping, water level recovery is measured for a significant period of time or until recovery occurs. If full recovery does not occur after a reasonable length of time (usually the elapsed time of pumping), further investigation regarding the water-supply-yield potential is warranted.

In deep aquifers, it may not be economically feasible to construct observation wells for test purposes. However, neighboring wells in the same aquifer may provide useful data.

Aquifer Modeling

The next phase of new well development is using aquifer modeling procedures to evaluate the sustainable yield, recharge potential, specific yield, and other factors pertinent to the development of a long-range water supply, which is projected to be costly. Many aquifer modeling systems and digital model programs are available for evaluating aquifers and well fields.

Aquifer modeling essentially began in the late 1950s and early 1960s with the development of the electric analog model. Both passive and active element models were used then. The active element models, using complicated and sophisticated electronic gadgetry, soon fell out of favor because of the high technical level of electronic capability and the high level of groundwater hydrology skills needed for their operation.

The passive element system, which consists of a simple resistor network scaled to the physical features of the aquifer for a constant transmissivity, still has some use in analyzing the steady-state yield of the flow system. Boundary conditions, such as negative boundary, and surface water infiltration sources, such as rivers, are modeled by the appropriate resistor networks. A voltage is applied to the resistor network with the current inputs varied on a trial-and-error basis until the voltage pattern approximately matches the observed water-level elevation for the aquifer area being modeled. When this calibration scheme has been completed, additional wells can be simulated with an immediate observation of the mutual interference or impact on the system. This method has been found useful to determine the steady-state yield of marginal aquifers and more recently to control the water-level elevation in known contamination areas for preventing or precluding contaminant migration into the active wells.

With the current development of high-capacity personal computers, aquifer models are now available to all practicing geohydrologists. Although many simplifying assumptions are used in the development of these programs, much useful information and insight into the interaction of wells can be obtained. If model data are calibrated to field experience, accurate predictions of future effects because of expansion or development of the well field can be achieved. Many specialized models also exist to analyze the aquifer characteristics as well as the movement of contaminants through the aquifer. Operational models may be developed to assist in the evaluation of day-to-day data for identifying any trends that require future maintenance and possible repair expense.

Groundwater Management Plan

In the aquifer modeling phase of the evaluation, a total groundwater management plan should be developed for the utility operating system.

The plan should include, but not be limited to, identification of the aquifer's sustainable yield, the major sources of recharge, and the magnitude of each source making up the sustainable yield projected. Multilevel aquifers may need to be evaluated separately. Any water quality differences associated with various sources of available recharge should be fully identified and their impact assessed as a result of future well field development. If installed pumping capacity is projected to exceed the sustainable yield, then a storage yield analysis of the aquifer should be made to determine the length of operating time available at the excess yield rate. The designer should consider the impact or availability of water rights and the impact of water withdrawals on other water users existing in the vicinity of the proposed development.

The study's findings should be made available to the facility's owner or operator and a disclosure made of any developmental or operational problems. Any limitations of the aquifer system should be identified.

Existing well fields. Procedures may change somewhat when investigating an existing well field to see if expansion is appropriate. In this situation, a review is made of the current operating history of the individual wells. If a significant lowering of static water level has occurred progressively through the years, it may be obvious that the aquifer is being pumped in excess of its sustainable yield. Expansion under these conditions may provide for a short-term water supply of increased capacity, but may not meet long-term water supply needs.

Often, well yields have declined dramatically in an existing well field requiring an expansion of supply, but little information is available on the water-level conditions in the well field. The basic cause may be the plugging of existing wells. Test drilling and monitoring well installation both in and out of the well field may be necessary to obtain the hydrologic data needed for appropriate evaluation.

The next phase of the expansion investigation program would be to drill test holes at prospective well sites to obtain information on the geologic formation, samples for laboratory evaluation, geophysical logs for well design detail, and a piezometer pipe or monitoring well, which would be installed to allow for water quality sampling.

Hydrologic data. Depending on the age and time of well field installation, hydrologic data may or may not be available to evaluate the formation hydraulic conductivity and storage coefficient values. If such data are not available, it may be possible to use one of the existing production wells and add the appropriate array of observation wells around this well to conduct a special constant-rate aquifer pumping test.

Similarly, the recovery of water levels should also be monitored when pumping is concluded. Operating wells adjacent to the test pumping well should be left in a constant mode before, during, and for several hours after the conclusion of the pumping test. The constant mode means that wells that need to operate should continue in a continuous operating mode, and those wells not needed should be shut off during the test period.

At the conclusion of this testing phase, an aquifer modeling program may be conducted to determine the effect of additional wells on the present operating characteristics of the well field. Also, it may be appropriate to develop a groundwater management plan for well field expansion.

Water-Level Monitoring System

Major well fields consisting of multiple wells should possess some type of continuous water-level monitoring system. Smaller well fields may have small-diameter monitoring wells capable of being read by hand periodically. Measurement frequency may vary from one week or less to several months between readings. All well fields should have water levels measured at least twice a year, usually in the early spring and late summer or early fall. For many utility systems, the early spring readings often represent the highest water levels during the year, while the late summer or fall readings represent the lowest water levels during the year after the summer period of heavy pumpage. In some industrial applications and some parts of the country, these trends may be reversed. New water-level measuring devices are now reaching the market to assist the utility operator in obtaining good water-level data for evaluation purposes.

WELL FIELD LOCATIONS

The location of a well field depends on a number of factors. The most important factor is water availability. Other factors include water quality, water temperature, and aquifer characteristics, including the storage and yield capabilities.

Availability of Water

The first and foremost consideration in well location is the availability of water and, in turn, the permeability of the aquifer. Obviously, the more permeable the aquifer, the better the water yield (in terms of gallons per minute obtained for the dollars expended). However, if an aquifer has somewhat similar water bearing characteristics through an extended area, then other considerations need to be made when determining where to locate specific wells.

One primary consideration (where the option exists) is to locate wells toward the recharge source. For example, when working with a stream-connected aquifer, the closer the wells are to the stream or river, the less drawdown and better well yield performance will be available.

A well placed near an aquifer edge, where an impermeable boundary or the discontinuation of the aquifer exists, will have greater drawdown and less potential recharge. In some shallow aquifers, there may be insufficient well depth to provide the hydraulic gradient necessary for obtaining the recharge available from a surface water source to meet the expected demand. Thus, wells located next to an aquifer barrier or too far from the recharge source may be limited in their potential long-term yield.

The primary recharge source of all fresh water obtained from the aquifer is precipitation. Rainfall, snowmelt, and the accumulation and melting of ice can have tremendous beneficial effects on determining the available water supply to individual wells.

Water Temperature

Wells located next to a body of surface water may experience a substantial temperature change in response to the degree of water infiltration from the surface water source. For example, a shallow stream or lake may have water warmed by solar radiation to 20°F or greater above normal with the water being directly filtrated to a well nearby. Thus, the groundwater temperature subject to rapid infiltration may vary considerably from that found in the initial investigation. However, if the wells are set back from the surface water infiltration source (a distance equivalent to six months travel time or greater), then little or no change in groundwater temperature should occur.

One beneficial use made of this groundwater-temperature-change phenomenon has been accomplished by surface water source plants developing groundwater in the vicinity of a river intake. During the freezing months, groundwater supplements the river water to maintain appropriate water temperatures in the treatment facility's basins, which prevents freezing and allows for proper chemical reactions. During the late summer months, it is possible to pump the supplemental wells and then infiltrate warmer-than-normal water from a river or surface source to the aquifer, which will remain dormant in the immediate vicinity of the wells during nonuse periods. Thus, it is possible to store solar-heated water in the aquifer adjacent to the stream for subsequent use during cold-weather months.

Water Quality

It may be important to evaluate the water quality of the recharge source. Special studies have indicated that brackish water may leak from certain bedrock aquifers as a result of large-scale pumping of a shallow interconnected alluvial aquifer. In this situation, relocating the wells closer to a freshwater recharge source or drilling wells to intercept the brackish water from the bedrock system may be required to control and properly manage groundwater quality in an area. In summary, with proper identification of the geologic and geochemical characteristics of the aquifer system, many operational alternatives may exist that would greatly enhance the economic operation of the utility system.

Before constructing a new well field consisting of multiple wells, a complete inventory and investigation of contamination sources needs to be made. Many different contamination sources exist that need to be inventoried.

Agriculture. Fertilizers, insecticides, and fungicides can have a particularly devastating effect on groundwater quality. For example, anhydrous ammonia (a common liquid fertilizer), readily turns to gas when vented to an atmosphere heavier than air. This gas can be easily redissolved by precipitation or local water sources. The nitrate from the fertilizer is trapped and can later move into the water table, which may result in nitrate contamination of the aquifer. Agricultural chemical spillage can also result in significant groundwater contamination problems.

Industry. Another major source of groundwater contamination has been motor fuel leakage from underground storage tanks. Consequently, motor fuel storage and distribution facilities in an area should be identified and monitored for potential groundwater contamination. Local ordinances often require motor fuel distributors to inventory their storage tanks frequently to detect leaks at the earliest possible time.

Other industrial facilities are major contaminators of groundwater. While much of this contamination is inadvertent because of normal materials handling procedures, contamination remains a serious problem. For instance, coal stockpiling for power plants can result in the leaching of phenols into the underlying aquifer. Salt for road deicing has been found to be a major groundwater contaminant if improperly stored and handled. Continued spillage of solvents on a cracked concrete floor will allow seepage of these materials into the underlying soils and eventually into the aquifers. Many solvents can easily penetrate clays previously thought to be protective barriers to pollution.

Wastewater facilities. Wastewater facilities may or may not present a potential for groundwater contamination. Older facilities with porous clay pipe or combination storm and sanitary sewers that carry industrial wastewater may distribute the contamination along the collection lines. Impervious wastewater collection lines with lift stations and watertight plant processing facilities should pose no threat to groundwater quality.

Disposal sites. Landfills, garbage dumps, and hazardous waste sites pose significant groundwater contamination risks. In addition, septic tank drain fields in aerobic soils may pose a contamination threat to an area. Septic tank drain fields in anaerobic soils may experience biological denitrification, resulting in little or no contamination risk to the groundwater.

Any water supplier contemplating development of a major groundwater source should identify all potential contamination areas. Buffer zones with a radius of several hundred feet should be established around each potential area, which may exclude any groundwater development.

Data analysis. The collection and analysis of water quality data is an important step in well design. Quality control procedures should be used to make certain that good data are being analyzed. Careful analysis of the data will provide insights on water quality characteristics at various depths in the aquifer system. Analysis will also provide information on the extent of treatment that may be required. These data and their analysis will be essential in obtaining the necessary permits for water withdrawal.

Aquifer Storage and Yield Capacity

In determining the number of wells needed to meet demand, consideration must be given to both the storage and yield capacity of the aquifer as well as the demands of the distribution network. If high seasonal peak demands are anticipated or high fire flows are required, it may be necessary only to install additional wells to meet the instantaneous demand. The installation of additional wells may require larger size conduction piping than normal, which may or may not be a significant cost factor. If wells can be placed throughout the distribution system, then no major increase in cost except for the well and pump installation is involved.

Unconsolidated sand and gravel aquifers possess tremendous storage capacity that need not be duplicated in the distribution system. However, if the water supply is being developed from a consolidated rock aquifer of low storage capacity, then system storage at ground level or elevated water storage may be required to meet peak water demands of the distribution system. Here, the maximum water supply can be extracted by pumping the wells at a slow constant rate for nearly continuous operation. Excess supply is placed into system storage.

If the well field is remote from the point of use, then it may be more economical to downsize the connecting piping to meet the peak three-day or peak monthly use and provide supplemental pumpage from tank storage at the point of use in the facility.

Appropriate consideration of both the characteristics of the aquifer and of the user's needs will result in the most economical water supply system possible.

WELL DESIGN

The gravel-pack well design, originally developed in the mid-1920s, has proven through the years to be a very reliable method for large-capacity well construction throughout the world. The gravel-pack method places a select washed silica gravel between the sand formation wall and the inner well screen openings to provide appropriate separation of water from sand. When properly designed, the well will pump a reasonable yield of water from the permeable material with good efficiency and with no excessive wear because of sand on the mechanical pumping equipment placed in the well. With the selected gravel-pack material sized to the finer formation material to be screened, appropriate performance from the well installation will be achieved.

Fluid Flow

When locating well fields, a thorough knowledge of fluid flow in the area is important, as is an understanding of aquifer characteristics.

Hydraulic conductivity. Hydraulic conductivity represents the fluid flow velocity through the porous media tested at the existing temperature. The hydraulic conductivity, listed in feet per day or centimetres per second, is established on unity gradient. It is physically possible to flow water through a formation at velocities in excess of those defined by the hydraulic conductivity value. After numerous years of study and observation, it has been concluded that hydraulic conductivity may represent the limit of laminar flow for the fluid viscosity through the porous media being tested.

Laminar flow. In practice, most large-capacity wells are pumped in the turbulent flow zone rather than in the laminar flow zone as usually assumed. A formula has been developed to identify the limit of laminar flow from a particular well bore given the hydraulic conductivity of the formation. This formula has been called the Natural Formation Yield formula. It is based on the concept that the flow rate is a product of the borehole area times the velocity of water flow from the well formation. The formula, which includes a constant to adjust to the selected units, is as follows:

$$Q_{nfy} = K_g \times L_s \times D_w / 5500 \qquad \text{(Eq 11-1)}$$

Where:

Q_{nfy} = the natural formation yield, in gallons per minute
K_g = the formation hydraulic conductivity, in gallons per day per square foot
L_s = the length of well screen or the thickness of the formation, whichever is less, in feet
D_w = the well bore diameter, in inches.

A designer who applies the above equation to existing wells or to new wells being designed may be surprised by the low yield calculated for the well under consideration. The yield obtained from this formula represents the limit of laminar flow for the formation hydraulic conductivity, thickness of the formation to be screened, and the borehole diameter to be drilled. The use of this equation assumes uniform flow through the column length of screen or formation.

Field experience has shown that the transitional phase of fluid flow is approximately 2.3 times the laminar flow rate. Field experience has also determined that the limit of turbulent flow that can be reasonably derived from the well is approximately 10 times the natural formation flow limit for laminar flow. The optimum point of operation for most wells is approximately twice the natural formation yield rate. The ideal rate is the limit of laminar flow, as defined by the natural formation yield for a well. At this rate, a significant improvement in well specific capacity can occur over that normally observed. There are occasions when this savings in energy would be quite significant.

Well screens. It has been assumed that water will enter a well screen uniformly through the column length available. Actually, when the pump suction terminates above the top of the screen, the water enters the top increment at the beginning of turbulent flow. The energy gradient is pushed downward by the frictional head created until the bottom of the screen is reached or the pump energy is dissipated.

Releasing the pump energy at the bottom of the screen will usually result in more uniform flow through the column length of screen. Hydraulic flow friction losses between the screen and pumping equipment must be checked before installation. If the deep-setting pump is used or the suction extended, a low-water-control device may be installed to prevent lowering the water level below the top of the well screen.

Material Selection

Selection of the gravel-pack material for a given formation has largely been left to the experience of the designer. The following are descriptions of various well conditions and the type of material selected for those conditions.

Ratios of the D_{50} size of gravel pack to formation have ranged from 4 to more than 12. However, the flow regime classification was not given nor considered in those testing procedures. It has been found that if the design yield of the well is at or near the natural formation yield, the gravel-pack size to formation ratio may range from 9 to 12, with the gravel-pack uniformity coefficient being approximately equal to that of the formation. In this instance, the selection of the gravel-pack material is essentially noncritical. The fact that a deep-setting pump or extended suction is required to achieve uniform laminar flow is of importance.

As the fluid-flow phase approaches the turbulent flow range, the ratio of the D_{50} size of gravel pack to the D_{50} of the sand may range from 6 to 9. Again, the uniformity coefficient of the gravel-pack material should be approximately equal to that of the sand formation. Gravel-pack selection should be of concern in this instance.

If the well is to be operated into the turbulent phase of fluid flow, the water quality and the uniformity coefficient of the sand formation plays an important role in the gravel-pack selection and the service life of the well. The uniformity coefficient is defined as the ratio of the D_{60} size to the D_{10} size for the material passing through the sieves used for particle-size analysis. If the sand formation has a uniformity coefficient of 2 or less, it is considered a uniform sand material. Therefore, the gravel-pack material selected should have an equally low uniformity coefficient, and the ratio of the D_{50} size to the formation size may range from 4 to 6.

If the water in the formation has a low pH and a low total dissolved solids content, then successful well operation may occur near the turbulent flow limit. As a uniformity coefficient increases in size, indicating a greater differential in particle size of the formation, the ratio of gravel pack to the formation must decrease. Ratios of the D_{50} gravel pack to the D_{50} of the sand may vary as low as three to four times that of the formation.

Similarly, the uniformity coefficient of the gravel pack should also increase in size, but may not increase to the same degree as the formation material. If a significant percentage of fines exists in the material (that is, 10 percent or more fines passing through a 70 mesh US sieve), the gravel-pack material may have to be sized exclusively to the fine end of the sand gradation. The coarse part of the formation may actually be coarser than some of the gravel-pack material used. The well will still function properly if the gravel-pack material is a well-washed select silica-type gravel material of good hydraulic conductivity. It should be noted that extended high-rate pumping of sand and gravel wells with a uniformity coefficient greater than about 5 may result in sand packing, thus reducing well yield.

Well Screen Selection

A well screen should normally retain between 85 and 90 percent of the gravel-pack material installed in the well. The screen should be as long as practical for the existing formation conditions. The hydraulic conductivity of most unconsolidated formations is usually much less in the vertical direction than in the horizontal direction. With this condition usually present, it is necessary to have the screen installed opposite all the formation present to produce the water yield expected.

Entrance velocity. Recently, controversy has developed regarding the allowable entrance velocity through a well screen. The practice has been to limit the velocity through the well screen to approximately 0.1 ft/s (0.03 m/s), or 6 ft/min (0.03 m/s). Recent laboratory tests show that this velocity can be increased. Extensive research, including mathematical analysis, review of practical experience, and extensive model testing, have established that entrance velocities up to 2.5 ft/s (1 m/s) may result in minimal head losses.

However, an important factor in determining actual entrance velocity is the "effective area of opening," which may be less than the measured opening or the manufacturer's published open area. A packing factor may develop around the outside of the well screen because of the particle size of the gravel pack and the shape of the screen opening. Sharp, angular openings usually result in less packing than smooth, flat openings. Blocking percentages may vary from less than 30 percent to more than 50 percent of the manufactured open area of the screen.

The well screen diameter and casing diameter should impose no frictional fluid losses in the up-hole flow. The fluid velocity should never be greater than 5 ft/s (1.5 m/s). The well screen and casing diameter selection is largely governed by the pump size requirements usually being greater than the upflow friction head requirements. The well designer is cautioned not to overlook the fundamental mechanics of fluid flow when considering flow to and around the pumping equipment to be installed in the well.

Screen length. The well screen length may be limited to the thickness of the geologic formation or formations if multilevel screens are used. In thick sand aquifers of good hydraulic conductivity, the minimum well screen length may be determined by the following formula:

$$L = \frac{Q}{A_e \, V_e \, (7.48)} \qquad \text{(Eq 11-2)}$$

Where:
- L = the length of screen, in feet
- Q = the quantity specified, in gallons per minute
- A_e = the effective aperture area of the screen, in square feet
- V_e = the design entrance velocity, in feet per minute.

The effective aperture area is obtained by decreasing the manufacturer's open area for the specific screen by the appropriate blockage factor. If the blockage factor is unknown, it is suggested that 50 percent be used.

Types of Well Installations

Many variations of the gravel-pack well design system are used today.

Straight screen and casing. The most common variation is a straight screen and casing of the same diameter of one-piece construction installed into a borehole that has been drilled at least 10 in. (250 mm) larger in diameter than the casing and screen diameter. Assuming the borehole is relatively straight and the casing and screen materials are installed concentric to the borehole, a nominal minimum of 5 in. (130 mm) of gravel-pack material will exist at all points around the screen. The normal minimum thickness of gravel pack is 5 in. (130 mm). The limit of gravel-pack thickness may vary with the well drilling method used.

Two-piece construction. Another common well installation is the two-piece construction method. In this installation, a large-diameter casing is installed in a borehole, usually 4 in. (100 mm) in diameter or greater than the selected screen diameter, to the top of the sand formation to be developed. This casing is then pressure cement grouted in place. The water-producing formation may be underreamed to a larger diameter than the casing, usually between 28 and 36 in. (710 and 910 mm) in diameter, and the screen is installed with the top of the screen casing overlapping the bottom of the outer casing. The minimum length of this lap pipe should be 50 ft (15 m) and artesian pressures may require as much as 90 ft (28 m) or more in some instances.

The annulus between the well screen and the borehole is then gravel-packed near the top of the lap pipe. The uplift forces on the submerged gravel material between the lap pipe and outer casing must be less than the friction head through the gravel pack and screen material. It is sometimes possible to vent the lap pipe just above the outer casing or to place a seal between the lap pipe and outer casing to prevent the uplift of granular material into the well and pump.

Tubular wells. Many times an inexpensive form of well construction is desired since the formation may be relatively uniform or the yield and flow from the well is in the laminar phase of fluid flow. A tubular-type well may be constructed where the borehole is only slightly larger than the diameter of the well screen and casing to be installed. The well is drilled and the casing and screen installed, which allows the natural formation to collapse around the well screen. Water is then pumped directly from the formation through the screen. In this situation, specially manufactured wire-wound well screens can be custom made to vary the slot width in accordance with the distribution of grain size of the sand formation. In large-capacity wells, it may not be practical to install a special custom-made stainless steel-type well screen designed for a specific well site and accomplish the work in a timely manner.

Construction techniques. In all gravel-pack well construction, a minimum of 20 ft (6 m) of surface casing should be grouted in place. (Surface-casing grouting requirements will vary from state to state.) This surface casing prevents the direct entry of surface water into the well. Wells constructed in consolidated rock aquifers usually have a casing extended through the overburden material into the top of firm rock and are pressure cement grouted in place. Open rock hole construction is then used to the bottom of the formation to be developed.

In limestone formations with solution channel development, sediment may be pumped for a considerable amount of time until the fluid phase of flow changes to laminar flow in the formation. The well depth in consolidated rock, such as firm sandstone, limestone, or dolomite, should penetrate all of the formation for best performance. The well diameter selection is largely governed by the size of pump to be installed in the well. In practice, open-hole wells are sometimes enlarged by acid treatment or explosion to increase the effective well diameter and fracturing of the well bore.

WELL DEVELOPMENT

Choosing the specific method of developing a well should be the responsibility of the well construction contractor and not the owner. However, the contractor's performance criteria can be designated by the owner. Methods typically used for well development are increasing rate pumpage in combination with surging, surging, high-rate pumping, use of explosives, high-velocity jetting, chemical agents, pressure acidizing, and hydraulic fracturing.

Increasing Rate Pumpage and Surging

Increasing rate pumpage in combination with surging is effective in developing wells in unconsolidated formations. Here the process is to first operate in the laminar flow range with a deep-setting pump or extended suction to recover as much of the drilling fluid lost to the formation as possible. When the discharge water begins to clear, gentle surging may assist in the development process. Pumping rates are increased in steps until the desired development is achieved.

Surging

Surging is probably the most frequently used development method. It may be used as part of the cleanup procedure after explosive work, in conjunction with chemical agents, or as the sole means of moving formation fines into the well bore.

In its simplest form, surging employs a plugger, or surge block, on the tool string, which is moved up and down in the well casing. This creates flow first in one direction and then the other at the well face. The result is that fine materials settle into the bottom of the well and then are bailed out. The surge block used may be solid and tight fitting or may be constructed with a check-valve-type arrangement. This arrangement gives a strong inward surge as it is raised in the well and a weaker backwash action as it is lowered. When the static water level is near the surface, the valved surge may actually pump the well while surging and thereby remove small suspended fines as they enter the well.

Surging may also be accomplished by using compressed air or alternate pumping and backwashing with water. With air, a provision is usually made to close in the top of the casing so that air pressure can be used to drive the water in the well bore out through the screen or bore face. Then, by proper valving, water can be rapidly pumped by airlift to remove fines. Air must not be driven into the formation because it may collect in pockets and reduce the aquifer production by air locking. Also, if the surging is too violent, a water–air mixture may tend to move upward along the outside of the well casing and rupture through to the surface. Such ruptures cause many problems and may result in loss of the well.

Surging with water can be accomplished by rapid on–off pump operation. The pump is turned on until water just starts to be discharged at the surface or into a tank, and then it is stopped, allowing the water to wash back down and out through the well face. After the cycle is repeated several times, the pump is then allowed to operate until the water is clear.

Backwash surging with chemical solutions has been applied with varying degrees of success. The most effective procedure requires a large tank for holding the mixed solution pumped from the well and adequate piping to permit rapid dumping back into the well when the pump is stopped. For the best results, the return rate to the well should be greater than the rate at which the well recovers.

High-Rate Pumping

When a well draws on a solid-rock aquifer, the development may be by simple pumping at a rate greater than the anticipated production rate to flush out the bore and adjacent aquifer materials. Although this procedure has been satisfactory in many consolidated formations, it is not usually adequate where a screen is used. However, it should be noted that regardless of what development procedures are used, some overpumping should be performed as the final step to clean the well and to demonstrate that performance will be satisfactory at a lower rate.

Use of Explosives

Explosives may be use in a variety of ways when developing wells. Perhaps the oldest use is to apply explosives to massive rock formations in order to fracture the rock. In this application, heavy charges are set off in the well bore at predetermined points, the objective being to radiate fractures outward from the bore, which will act as conduits. This technique has been quite successful under proper circumstances for consolidated rock wells.

Another common technique is to shoot dynamite (approximately 1 lb/ft [0.5 kg/0.3 m of hole]) throughout the most permeable portions of a formation. The purpose is to spall the bore face, which may be partially plugged with mud or other foreign materials.

Another method, which is used for both open-rock holes and screened wells, is the use of vibratory explosives. In this method, very light charges set off an extended series of explosions to produce intense, though not violent, vibrations that agitate the materials surrounding the bore. Sonic jetting can then be used to remove scale from screens.

For each of these explosive methods, a well rig must be available to bail out and clean the well. Where heavy charges are used, it is often necessary to break up the large pieces before they can be removed by bailing.

High-Velocity Jetting

The most recently improved method of well development is high-velocity jetting. For this method, a jetting tool is used to concentrate a fluid jet stream on the face of the borehole and penetrate a short distance into the aquifer. Fluid leaves the jet nozzle with a velocity of at least 150 ft/s (40 m/s). When this force strikes the formation, it breaks down dense compacted materials, rotary drilling muds, and even some chemical precipitates, and rather violent localized agitation ensues. Fine materials are dispersed in solution, and these materials can be purged by pumping the well at the time jetting is taking place.

The jets are rotated at elevations from 6 to 12 in. (150 to 300 mm) apart throughout the water-entry area of the well. Several passes are made vertically throughout the entry section until the pump discharge is clean. Improper use of jetting may result in loss of porosity at the well face and loss of well efficiency.

The rate of pumping is 15 to 20 percent greater than the rate at which fluid is applied at the jets, which keeps water in the formation moving toward the well. Pumping is easily carried out by the airlift method, if the static level is near the surface, or by a suction-lift-type pump.

Chemical Agents

Many formations respond well to common development methods if a chemical additive is used. For example, in the jetting method, chemical additives are often used where silts, clays, or drilling muds are known to be present. Wetting agents used with hexametaphosphates in concentrations of 5–6 lb/100 gal (2–3 kg/378.5 L) of jetting fluid have been shown to be particularly effective.

Acids may be used on iron and carbonate deposits to disperse them and thus facilitate their removal. Wetting agents and hexametaphosphates are employed to disperse colloidal deposits, as just discussed. In well reconditioning, where biological or chemical precipitation has resulted in plugging, chlorine and acids have been used with considerable success.

Pressure Acidizing

Pressure acidizing is characterized by injecting acid at high concentrations into a formation at high pressures. The result is the injection of large volumes of solution and increased radial penetration from the well bore. The solution used is commonly inhibited muriatic acid, which is supplemented with gelling agents to modify its flow properties and retarders to slow down reactions and ensure that at least some of the high-strength acid will penetrate the extremities of the formation. This method has application only in rock aquifers where a good isolation or seal from overlying formation exists. Its success is keyed to the relatively great radial penetration.

Hydraulic Fracturing

Hydraulic fracturing consists of injecting into the well fluid that is under pressures great enough, approximately 1 psi (7 kPa) for each foot of well depth, to actually separate the aquifer along bedding planes and fractures. Gelling agents and selected sand are added to the fluid in order to reduce the required fluid volume and hold the partings open after pressure is removed. Little is known about the size of the openings created, but increases in apparent hydraulic conductivity of up to 20,000 gpd/ft^2 (75,000 L/d/m^2) indicate they must extend to considerable lateral distances.

As in the case of pressure acidizing, this method can be applied only to rock aquifers, and the well casing must be firmly sealed and anchored in place.

Well Performance Criteria

Responsibility. If the design engineer specifies, in detail, the exact size and nature of material to be installed in the well and the well does not achieve the desired performance criteria, the design engineer may suffer some of the responsibility for failure of the system. Consequently, the design engineer may want to specify only the general conditions and level of results and performance to be achieved, leaving the full responsibility of accomplishing these results to the contractor.

Allowable sand. In designating the amount of sand allowable in well production, end use of the well water and treatment systems should be considered. If the water is to be pumped directly to the distribution system with no filtration or other conditioning, then the amount of allowable sand may be as low as 1 ppm from pump startup. Normal operation to a water treatment facility may allow 3–5 ppm of sand from startup. Less critical uses may allow as much as 10 ppm 15 min after pump startup as acceptable. Any level specified can be achieved with proper design and installation.

Well development after completion of construction in a sand formation may require the contractor to perform extensive work to achieve the level of sand production and yield specified. Complete development may be determined by sand content of the pump discharge water. Sand content should average not more than 5 mg/L for a complete pumping cycle of 2 h duration. No less than 10 measurements should be taken at equal intervals of time to permit plotting of sand content as a function of time and production rate. There should be no appreciable increase in specific capacity during the last several hours of pumping and surging.

Specific capacity. Another performance criterion that can be achieved by the well contractor is to develop the well to a minimum specified percentage of the theoretical specific capacity. Theoretical specific capacity may be determined from the transmissivity—a number obtained from preliminary time-drawdown data or from the distance-drawdown data (if available). Usually the time interval used to obtain time-drawdown data is from approximately 10 min to approximately 200 min; this best represents the flow of water from the formation material into the well. Longer term transmissivity data may be influenced by negative or positive boundary conditions, leakage, delayed yield from storage, and other factors varying from the assumed conditions on which the formulas are based. The storage coefficient used in this calculation may be somewhat smaller than the long-term value usually associated with the formation conditions. Experienced judgment must be used and some latitude allowed to the well contractor in performing under this condition.

Testing. There is no set time period for well development in which to accomplish the required results. An inspector or owner's representative should observe closely the character and quality of fine materials being removed from the well and the procedures being used by the contractor during development. The inspector should perform all acceptance testing procedures using a consistent procedure that is acceptable to both the owner and the contractor. At the conclusion of the development procedure used by the contractor, a temporary contractor-furnished test pump should be installed in the well for pump-testing purposes.

WELL TESTING

Once a well has been constructed and developed, it should be tested to determine its capacity and other important characteristics.

Pump Specifications

A test pump that is installed in a well must be capable of pumping continuously for 100 h or more at acceptable drawdowns without breaking suction for the conditions encountered. The owner should specify the maximum capacity expected to be test pumped from the well and the maximum pumping level depth, which sets the size and horsepower of the equipment to be used for testing purposes. It is often desirable to add an extended length of suction pipe in deep wells to extend the suction near the bottom of the well screen. This long suction pipe transfers the energy of the prime mover to the bottom of the well screen, which results in a more uniform flow through the entire length of the well screen installed. This may be important in development of bottom screens for multilevel screened wells.

Nearly any motive power that is reliable may be used for a deep-well-turbine test pump. If the testing period is set, it is desirable that no interruption occur during the test. The driver should also have sufficient power to drive the pump at the rate of speed required for full pump capacity to be developed. If a constant-speed electric motor is used, the rate adjustment can be accomplished by adjusting a throttling valve on the pump discharge.

Horsepower. The horsepower required for pumping will be

$$\text{HP} = \frac{Q \times H_f}{3960 \times E_p} \text{ (for water)} \quad \text{(Eq 11-3)}$$

Where:
- HP = the horsepower required to drive the pump
- Q = the water to be pumped, in gallons per minute
- H_f = the total head or lift of water to be pumped, including column friction losses
- E_p = the pump efficiency (assumed at 0.70 if actual pump bowl efficiency is unknown).

The weight of water is normally considered to be 8.34 lb/gal (1.00 kg/L). However, drilling fluids can exceed 9 lb/gal (1.08 kg/L). If a well is producing gas for some reason, the weight of water may decrease to less than 8 lb/gal (0.96 kg/L). Note that these factors may cause some difficulty in the development or operation of the well or initial pumping equipment.

Pumping Tests

A pumping test is used to compare the performance of the constructed well with that anticipated and to establish final specifications for pumping equipment to be permanently installed on the well. The pumping test consists of pumping the well for a long period and observing water-level changes in nearby monitoring wells during the pumping period and during recovery after pumping ceases.

Before beginning any pumping test, some rest period should exist between the development work and prior pumping and that of the test itself. The minimum time should be considered 4 h, but 12–24 h is preferred to allow for full recovery.

Flow rate. A suitable measuring device should be installed on the test pump discharge to verify the flow rate. The orifice-plate method is commonly used. Other methods, such as pumping through a weir box, measuring open-pipe discharge, or installing a calibrated flowmeter, are acceptable.

Orifice plate. Two types of orifice-plate installations are used for measuring flow rate. In one method the orifice is installed in the line, and in the other method the orifice is installed at the end of the discharge pipe. For an orifice installed in a pipeline other than at the discharge end, the flow may be calculated by the following formula:

$$Q = 448.83 Ca \sqrt{\frac{2gH}{1 - R}} \qquad \text{(Eq 11-4)}$$

Where:

Q = the flow, in gallons per minute
C = the orifice coefficient (0.61 for a thin sharp-edged orifice)
a = the area of the orifice, in square feet
g = the acceleration of gravity (32.2 ft/s^2)
H = the difference in pressure between a point in a pipeline that is one pipe diameter upstream from the orifice and a point that is one orifice diameter downstream from the orifice, as measured by a manometer, in feet
R = ratio of inside pipe diameter to the orifice diameter.

If an orifice is used at the discharge end of a pipe, the flow is equal to

$$Q = 8.03 ka \sqrt{H} \qquad \text{(Eq 11-5)}$$

Where:

Q = the flow, in gallons per minute
k = the flow coefficient, as given in Table 11-1
a = the area of the orifice, in square inches
H = the pressure in the pipe at a point 2 feet (0.6 m) ahead of the orifice, in inches of water above the centerline of the pipe.

Table 11-1 Values of k in Orifice Formula

Ratio of Orifice to Pipe Diameter	k
0.3	0.552
0.4	0.564
0.5	0.583
0.6	0.613
0.7	0.662
0.8	0.745

Weirbox. When pumping is done by airlift methods, a weir box is the most accurate means of measuring the well discharge. Several different weir types are used, such as the rectangular weir, the V-notch weir, and the trapezoidal weir. Weirs are placed at the end of a rectangular box or a channel, which allows full dissipation of the air from the water before measurement. The simplest form of weir is the 90° V-notch weir, which is suitable for most well testing work. The flow for this weir type is calculated as follows:

$$Q = 2.28 \, H^{5/2} \qquad \text{(Eq 11-6)}$$

Where:

Q = the flow, in gallons per minute
H = the height of water above the bottom of the weir notch, in inches.

Open-pipe discharge. The open-pipe discharge method is generally based on the free fall of water being discharged from a level pipe (Figure 11-3). The pipe must be flowing full, and the method is limited in the range of ratos that can be measured. This method is usually about 90 percent accurate. In method A, both the free fall and distance to the point is calculated using the following formula:

$$Q = \frac{2.45 D^2 x}{\sqrt{\frac{2Y}{32.16}}} \qquad \text{(Eq 11-7)}$$

Where:

Q = the flow, in gallons per minute
D = the inside pipe diameter, in inches
x = the horizontal distance, in feet
Y = the vertical distance of fall, in feet.

To simplify the equation further, it is possible to measure along the top of the flowing stream so that the vertical fall is always 1 ft (0.3 m) and to measure the horizontal distance x in inches as shown in Figure 11-3 for method B as follows:

$$Q = 0.813 \, D^2 x \qquad \text{(Eq 11-8)}$$

Flow meter. A satisfactory method used to measure the flow of water is a commercially available water meter. Water meters should be periodically tested to verify their accuracy. Many water meters have a flow-rate indicator and a totalizing dial

142 GROUNDWATER

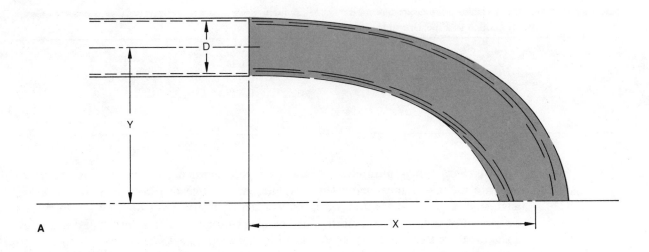

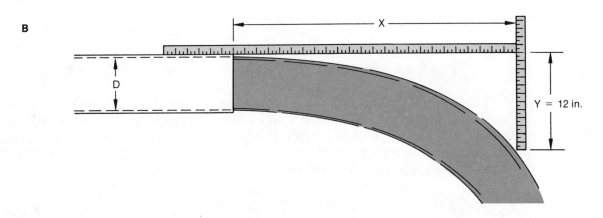

A—*D* is inside pipe diameter, in inches; *X* is horizontal distance, in feet; *Y* is vertical distance, in feet. *B*—*x* is in inches; *Y* is always 12 in.

Figure 11-3 Open-pipe discharge method for measuring pumping rate.

that may be read periodically to verify the average flow rate of the pump test. A flow-velocity meter or a Pitot tube may also be used if the cross-sectional area of the pipe is known. Regardless of the flow measuring device used, it is important to record the well discharge rate at periodic intervals so that any variation may be noted and averaged for analysis.

Depth to water level. An accurate device for measuring depth to water level must be used in conducting the pumping test. A steel tape with a chalk marking system may be used to determine the water level accurately, although this is not always feasible inside of a pumping well. An alternative method for measuring the water level in wells is an electric-line water-well sounder. With the sounder, a meter, either the milliamp or millivolt type or light, is activated when contact with the water surface occurs. Air-line devices can be used where precise measurements are not required. Air-line devices usually have an accuracy of ±1 ft (0.3 m). New electronic transducers are now available to measure water levels accurately. These transducers may be coupled to a computer system for recording measurements at frequent intervals. It is normally desirable to have measurement accuracy within 1 in. (25 mm) tolerance to analyze the data from a test pumping well. If observation wells are available, the drawdown accuracy desired is ±0.1 in. (3 mm).

Well efficiency and flow characteristics. An increasing rate step drawdown test is used to determine well efficiency and flow characteristics. The test starts at a slow rate of pumping, usually 25–30 percent of the design yield of the well, and pumped for 1–3 h. The rate may then be increased to 50 or 60 percent of the design yield of the well and pumping continued for another time period. The third pumping step may be the design flow rate for the well, and a fourth step at 115–130 percent of design yield may be added if desired.

At the completion of the test pumping, recovery of water levels should be measured for approximately 2–4 h. In the first step at the low rate, flow should be in the laminar flow phase that often gives the highest specific capacity with little or no sand pumping. As the pumping is increased for a normal well, a small decrease in well specific capacity will occur because of the increase in frictional fluid losses developing around the well bore. If the well specific capacity increases with an increase in pumping rate, it would be questionable whether the well was fully developed and that an additional development was occurring during the pumping test. If a significant decrease in well specific capacity occurs with increased pumping rate, it would be questionable whether the well efficiency had been achieved. Further development may be warranted or plugging of an existing well may be occurring.

A rough estimate of well efficiency may be achieved by dividing the high-rate specific capacity by the low-rate specific capacity.

Aquifer flow characteristics. A constant-rate pumping test of 24 h or more duration is usually conducted to determine the aquifer flow characteristics of transmissivity and storage coefficient for a newly constructed well. Normally the well is rested a minimum of 12 h before start of the test. No interruption is allowed in the pumping test during the first 8 h of pumping. Thereafter, an interruption of 1 percent of the elapsed pumping time can be tolerated if the test is extended by three times the interruption interval. At the conclusion of pumping, recovery should be made for a minimum of 4 h or until full recovery to the observed static water level is achieved.

Water-level drawdown measurements are made at frequent intervals during the early portion of the test in accordance with procedures previously outlined. It is suggested that with shallow wells 200-ft (60-m) deep or less that two or more observation wells or piezometers fully penetrating the aquifer be installed to observe the drawdown in the formation.

Data analysis should identify the degree of well development (satisfactory or unsatisfactory), the short-term transmissivity of the aquifer, the long-term transmissivity of the aquifer for future yield projections, the storage coefficient of the aquifer during the short-term test, and the projected storage coefficient or aquifer specific yield for long-term yield projections. Other parameters that should be identified are the efficiency of the water supply well, the apparent safe yield of the well, the radius of influence of the well, the projected well interference on drawdown of other wells (if appropriate), the projected pumping level in the water supply well for the design discharge rate, and the recommended capacity of the permanent pumping unit for the water supply well. The feasibility of the proposed spacing between wells or recommended minimum spacing between wells and their capacities should be determined and used in the design of the well field or its future expansion.

WELL OPERATION AND MAINTENANCE

After all pumping equipment is installed and the housing facility is connected to the water supply distribution system, the well condition is often forgotten. Well operation and maintenance procedures should include periodic inspection of the pumping equipment and the condition of the well itself.

Inspection

A simple monthly test of well performance that should be conducted is to observe the standing or static water level after a short shutdown period, and then observe the pumping water level at a specific interval of time following pump startup. For example, a given well in a well field may be shut off the night before and allowed to recover overnight and then the standing water level measured before pump startup. The pump is started, the pressure is observed, the discharge capacity is observed, and the pumping water level is observed 1–4 h after startup. The time interval should be consistent for every measurement.

Analysis. Data from this test may be represented by a vertical line, with the top of the line representing the depth to water level or water-level elevation for the nonpumping condition, and the bottom of the vertical line representing the depth to water level or pumping-level elevation for the pumping rate. If the line remains the same length for several years, the specific capacity is remaining constant, which normally should occur. If the line begins to lengthen, specific capacity is decreasing and further inspection may be warranted. If this line becomes shorter, the pumping rate may be decreasing because of some problem.

Normally, the static water level will show seasonal fluctuations, dropping during high-use periods and recovering to a normal level during low-use periods. If the static water-level trend is cycling downward, overdraft of the aquifer may be occurring. Regular monitoring of the well operation will give early warning to developing problems with the pumping equipment or the well.

Water Demand

The maximum service life with the greatest water volume derived from an aquifer is obtained by pumping at a slow, constant rate. If water demands are cyclic, it may be more appropriate to install additional wells in close proximity to the existing wells to meet the peaking capacity of the distribution system. Additional wells, closely spaced, can use the storage capacity of an unconsolidated aquifer to meet short-term peak demands. There is no intent in this situation to continuously pump all wells unless appropriate studies have been made to verify the sustained aquifer yield. If the water supply demand requires nearly constant pumping, proper spacing between wells is required to use the effective recharge of the aquifer most efficiently.

Closely spaced wells in consolidated rock aquifers with fracture band characteristics can quickly develop adverse mutual interference effects. In these aquifer types, a few appropriately spaced wells can provide all of the yield available from the aquifer system. Additional wells in the same fracture system only decrease the yield from the existing wells. Supplemental distribution system storage may be required to meet peaking demands in this situation.

Well Failure

Many individual well failures occur, but seldom is there a failure of the aquifer system. Well failures may be grouped into three principal categories, which are as follows: mechanical failure, chemically related failure, and biological plugging.

Mechanical failure. Mechanical failures occur when the well casing or screen collapses or corrodes through. This is a structural material failure of the well.

Another type of mechanical failure that often occurs in sand aquifers with a high uniformity coefficient is mechanical plugging. Plugging is caused by the migration of fines as a result of turbulent flow in the aquifer. If the gravel-pack material is properly sized to restrain the migration of fines, the fine particles will actually reduce the pore space and the hydraulic conductivity of the formation immediately surrounding the gravel-pack material. Well recovery from this type of plugging is extremely difficult. Prevention by appropriate design and pump operation is the best remedy.

Chemically related failures. Chemically related failures can occur with the interflow of waters of different quality from one aquifer level to another. A chemical precipitate may form at the face of the aquifer in one section, resulting in plugging and diminution of well yield with increased drawdown.

Another chemically related failure is electrolysis or corrosion because of external forces entering the well or electrical systems being grounded through the pump to the well. Large holes can develop quickly through the casing material or well screen by an induced voltage following the distribution piping to the well. Occasionally, the water in the aquifer will contain excess dissolved carbon dioxide in a high hardness water that will plate out calcium deposits in the gravel pack and on the screen. This can usually be reversed by periodic treatment with an acid solution.

Biological plugging. Most aquifers are relatively free of bacteria. However, with the cyclic action of pumping equipment, it is possible to induce bacteria through the air vent in the base of the pump into the well. A well can easily be contaminated during construction if not properly disinfected. A well can occasionally become contaminated by the inducement of surface water through the aquifer if sufficient hydraulic conductivity exists.

Iron bacteria. The most common form of bacteria found in wells is a form of iron bacteria. These bacteria can survive in the slightly aerobic water, which contains approximately 0.5–2.5 mg/L dissolved oxygen. Some iron content in the water is also needed to support bacterial growth. There often is sufficient agitation with the starting and stopping of a pump in a well to provide the small amount of dissolved oxygen necessary for iron bacteria to grow. Although their numbers are small at first, iron bacteria eventually will plug the gravel pack and screen in the well vicinity. Normal chemical cleaning is often ineffective in restoring a well that has become badly contaminated with iron bacteria.

Slime-forming bacteria. Similarly, wells have been found to be contaminated with one or more forms of the slime-forming group of bacteria. In individual cases, these bacteria have been found to exist as far as 15 ft (5 m) or more from the well in the aquifer. These bacteria types often occur in shallow aquifers that are connected to a surface water recharge source. These bacteria are common in soil and can be induced through the pump base to become established in the well. Special sterilization techniques are often required to recover a well that has been badly plugged with this bacteria.

Disinfection and Maintenance

It is recommended after wells have been constructed and placed in service that they be periodically sterilized with chlorine solution to prevent bacteriological contamination. Well seals, pump base seals, well vents, pump bases, and other appurtenances to a well installation must be maintained in good repair if the well is to provide long-term, trouble-free service.

Bibliography

ARON, G. & SCOTT, V.H. Simplified Solutions for Decreasing Flow in Wells. *Proc. Amer. Soc. Civ. Engrg.*, 91 (HY 5):1 (1965).

BENTALL, R. Methods of Determining Permeability, Transmissibility, and Drawdown. US Geol. Surv. Water-Supply Paper 1536-I, 243 (1963).

BOULTON, N.S. Analysis of Data from Non-Equilibrium Pumping Tests Allowing for Delayed Yield From Storage. *Proc. Inst. Civ. Engrg.*, 26:469 (1963).

BOULTON, N.S. & STRELTSOVA, T.D. New Equations for Determining the Formation Constants of an Aquifer From Pumping Test Data. *Wtr. Resour. Res.*, 11:148 (1975).

CAMPBELL, M.D. & LEHR, J.H. *Water Well Technology.* McGraw-Hill Book Co., New York (1973).

CARR, P.A. & VAN DER KAMP, G.S. Determining Aquifer Characteristics by the Tidal Method. *Wtr. Resour. Res.*, 5:1023 (1969).

CHOW, V.T. On the Determination of Transmissivity and Storage Coefficients From Pumping Test Data. *Trans. Amer. Geoph. Union*, 33:397 (1952).

COLLINS, W.D. Temperature of Water Available for Industrial Use in the United States. US Geol. Surv. Water-Supply Paper 520-F (1925).

COOPER, H.H. JR.; BREDEHOEFT, J.D.; & PAPADOPULOS, I.S. Response of a Finite-Diameter Well to an Instantaneous Charge of Water. *Wtr. Resour. Res.*, 3:263 (1967).

COOPER, H.H. JR. & JACOB, C.E. A Generalized Graphical Method for Evaluating Formation Constants and Summarizing Well-Field History. *Trans. Amer. Geoph. Union*, 27:526 (1946).

DAGAN, G. A Method of Determining the Permeability and Effective Porosity of Unconfined Anisotropic Aquifers. *Wtr. Resour. Res.*, 3:1059 (1967).

DRISCOLL, F.G. *Groundwater and Wells.* Johnson Division, St. Paul, Minn. (1986).

EHLIG, C. & HALEPASKA, J.C. A Numerical Study of Confined-Unconfined Aquifers Including Effects of Delayed Yield and Leakage. *Wtr. Resour. Res.*, 12:1175 (1976).

ELLIS, A.J. The Divining Rod, A History of Water Witching (with a bibliography). US Geol. Surv. Water-Supply Paper 416 (1917).

Environmental Quality—1979, The Tenth Annual Report of the Council. US Council on Envir. Qual. US Govt. Printing Ofce. (1980).

FERRIS, J.G. Cyclic Fluctuations of Water Level as a Basis for Determining Aquifer Transmissibility. *Assemblée Generale de Bruxelles, Assoc. Int. Hydrol. Sci.*, 2:148 (1951). [Also in US Geol. Surv. Water-Supply Paper 1536-E (1962); and 1536-I (1963).]

FERRIS, J.G. ET AL. Theory of Aquifer Tests. US Geol. Surv. Water-Supply Paper 1536-E (1962).

Groundwater Manual—A Water Resources Technical Publication. US Dept. Inter., Washington, D.C. (1977).

HANTUSH, M.S. Analysis of Data From Pumping Tests in Leaky Aquifers. *Trans. Amer. Geoph. Union*, 37:702 (1956).

———. Nonsteady Flow to Flowing Wells in Leaky Aquifer. *Jour. Geoph. Res.*, 64:1043 (1959a).

———. Analysis of Data From Pumping Wells Near a River. *Jour. Geoph. Res.*, 64:1921 (1959b).

———. "Hydraulics of Wells." In *Advances in Hydroscience,* vol. 1, edited by V.T. Chow. Academic Press, New York and London (1964a).

———. Drawdown Around Wells of Variable Discharge. *Jour. Geoph. Res.*, 69:4221 (1964b).

———. Anaylsis of Data From Pumping Tests in Anisotropic Aquifers. *Jour. Geoph. Res.*, 71:421 (1966).

HANTUSH, M.S. & JACOB, C.E. Non-Steady Radial Flow in an Infinite Leaky Aquifer. *Amer. Geoph. Union*, 36:95 (1955).

HANTUSH, M.S. & THOMAS, R.G. A Method for Analyzing a Drawdown Test in Anisotropic Aquifers. *Wtr. Resour. Res.*, 2:281 (1966).

HELWEG, O.J.; SCOTT, V.H.; & SCALMANINI, J.C. *Improving Well and Pump Efficiency.* AWWA, Denver, Colo. (1983).

KROSZYNSKI, U.I. & DAGAN, G. Well Pumping in Confined Aquifers: The Influence of the Unsaturated Zone. *Wtr. Resour. Res.*, 11:479 (1975).

KRUSEMAN, G.P. & DE RIDDER, N.A. Analysis and Evaluation of Pumping Test Data. Bull. 11, Intl. Inst. Land Reclam. and Improve., Wageningen, The Netherlands (1970).

LANG, S.M. Methods for Determining the Proper Spacing of Wells in Artesian Aquifers. US Geol. Surv. Water-Supply Paper 1545-B (1961).

LOHMAN, S.W. Groundwater Hydraulics. US Geol. Surv. Prof. Paper 708 (1972).

MEINZER, O.E. The Occurrence of Groundwater in the United States, with a Discussion of Principles. US Geol. Surv. Water-Supply Paper 489 (1923).

MILLER, D.W., ed. *Waste Disposal Effects on Groundwater.* Premier Press, Berkeley, Calif. (1980).

MOENCH, A.F. & PRICKETT, T.A. Radial Flow in an Infinite Aquifer Undergoing Conversion From Artesian to Water Table Conditions. *Wtr. Resour. Res.*, 8:494 (1972).

NEUMAN, S.P. Theory of Flow in Unconfined Aquifers Considering Delayed Response of the Water Table. *Wtr. Resour. Res.*, 8:1031 (1972).

———. Analysis of Pumping Test Data From Anisotropic Unconfined Aquifers Considering Delayed Gravity Response. *Wtr. Resour. Res.*, 11:329 (1975).

NEUMAN, S.P. & WITHERSPOON, P.A. Theory of Flow in a Confined Two-Aquifer System. *Wtr. Resour. Res.*, 5:803 (1969).

PAPADOPULOS, I.S. & COOPER, H.H. JR. Drawdown in a Well of Large Diameter. *Wtr. Resour. Res.*, 3:241 (1967).

PINDER, G.F.; BREDEHOEFT, J.D.; & COOPER, H.H. JR. Determination of Aquifer Diffusivity From Aquifer Response to Fluctuations in River Stage. *Wtr. Resour. Res.*, 5:850 (1969).

PRICKETT, T.A. Type-Curve Solution to Aquifer Tests Under Water-Table Conditions. *Ground Water*, 3 (3):5 (1965).

PYE, V.I.; PATRICK, R.; & QUARLES, J. *Groundwater Contamination in the United States.* Univ. of Pennsylvania Press, Philadelphia, Pa. (1983).

Standard Methods for the Examination of Water and Wastewater. APHA, AWWA, and WPCF, Washington, D.C. (16th ed., 1984).

STERNBERG, Y.M. Transmissibility Determination From Variable Discharge Pumping Tests. *Ground Water*, 5(4):27 (1967).

THEIS, C.V. The Relation Between the Lowering of the Piezometric Surface and the Rate and Duration of Discharge of a Well Using Groundwater Storage. *Trans. Amer. Geoph. Union*, 16:519 (1935).

TODD, D.K. *Groundwater Hydrology.* John Wiley & Sons, New York (1980).

WALTON, W.C. Selected Analytical Methods for Well and Aquifer Evaluation. Illinois State Water Survey Bull. 49 (1962).

WEEKS, E.P. Determining the Ratio of Horizontal to Vertical Permeability by Aquifer-Test Analysis. *Wtr. Resour. Res.*, 5:196 (1969).

Index

Acids, 137
Acoustic logging, 31
Aeration, 115–16
Agriculture, 130
Airlift pumps, 90–91
Air-line devices, 142
Analysis, 33–34, 57–59
Aquifer response models, 71–73
Aquifers, 3–4
  boundaries, 62–63
  characteristics, 4–9, 122–25
  flow characteristics, 143
  isotropic vs anisotropic, 12
  modeling, 127
  storage, 131
  testing, 31
Artesian aquifers, 4
Artesian wells, 4
AWWA Standard A100-84, 48

Backwashing, 118
  surging, 136
Bacteria, 145
Bail-down method, 46
Biological plugging, 145
Bored wells, 40
Borehole geophysical logging, 29–31
Breaking suction, 99–101

Cable-tool drilling method, 42
Calcium hypochlorite, 54, 119
California method, 44
Caliper logging, 29
Capacity, 76
Capillarity, 13–14
Capillary fringe, 3
Capillary gradient, 14
Casing log, 31
Cementing, 49–50, 53
Centrifugal pumps, 81, 84–85
Check valves, 97
Chloramines, 118
Chlorination, 118–19
Chlorine, 53, 103, 118–19
Chlorine gas, 119
Chlorine residuals, 118
Coefficient of transmissivity
  See Transmissivity
Compressibility, 19–20
Cone of depression, 124–25
Confined aquifers, 4, 59
Confining beds, 3
Contamination
  See Water quality and contamination

Darcy's law, 14, 16
Data manipulation models, 71, 75
Deep-well turbine pumps, 85
Deformation models, 73
Diffuser-type pumps, 84
Direct-current resistivity method, 25

Disinfection, 53–54, 145
Distance-drawdown method, 61
Documentation, 36–37
Drawdown method, 59–63
Drill hole, 42
Drilled wells, 41–45
Drilling permits, 35
Driven wells, 40–41
Dug wells, 39–40
Dump-bailer method, 50
Dynamic head, 76

Elastic wave propagation, 27
Electric motors
  selection, 96
Electric tape, 56
Electrical resistivity, 25–27
Electrical resistivity logging, 30
Electrical sounding, 26
Exploratory drilling, 31–32
Explosives, 104, 137

Field coefficient of permeability
  See Hydraulic conductivity
Filtration, 117–18
Float-actuated recording devices, 56
Flow models, 72–73
Flow rate, 140–42
Flow meter, 141–42
Fluoridation, 119
Formation recognition, 42–43
Friction head, 76, 78, 79

Gamma-gamma logs, 31
Gamma logs, 31
Geophysical logs, 32
Geophysical surveys, 125
Glassy phosphates, 103
Granular activated carbon (GAC)
  treatment, 118
Gravel envelope method, 46
Gravel-pack well design, 131
  material selection, 133
  types, 134–35
Gravel-wall wells, 45–46
Gravels, 123
Gravity filters, 117–18
Ground-penetrating radar, 29
Groundwater
  demand for, vi
  information sources, 23–24
  locating supplies, 24–33
  management, 127–28
  movement, 8–9, 11–19
  occurrence, 2–4
  regional supplies, 21–37
  temperature, 109–110
  US importance, vi–vii
  See also Modeling; Treatment;
    Water quality and contamination
Grout, 49–50

Hardness, 111
Hazardous waste sites, 130–31
Head loss, 78
Head/submergence (H/S) ratio, 90
Heat transport models, 73
Hexametaphosphates, 137
Horizontal profiling, 26
Horsepower, 139–40
Hydraulic conductivity, 11–13, 15, 125, 132
   rock, 12–13
Hydraulic fracturing, 138
Hydraulic gradient, 8–9
Hydrochloric acid, 103
Hydrogen ion concentration (pH), 109
Hydrograph, 60
Hydrologic cycle, vi, 1–2
Hypochlorite ion, 118
Hypochlorous acid, 118

Impermeable-barrier effect, 62–63
Impulse pumps, 90–91
Incrustation, 101–104
Infiltration
   rates, 1–2
Inorganic salts, 22
Intermediate zone, 3
Internal storage, 123
Ion-exchange process, 117
Iron bacteria, 145

Jet pumps, 91
Jetting, 44–45
   high-velocity, 137

Laminar flow, 132
Land costs, 35
Land subsidence, 19–20
Land use, 22–23
Landfills, 130–31
Leaky aquifers, 59
Lime–soda ash process, 116–17
Lithologic logging, 32
Log suites, 31

Management
   groundwater, 127–128
   water quality and contamination, 114
Mass transport models, 73
Mineralization, 21–22
Mixed-flow pumps, 87
Modeling, 23
   aquifers, 127
   types, 71–75
   usage, 70–71
Monitoring
   water level, 128–29
   water quality and contamination, 33–35
Monitoring wells, 23, 33

National Electrical Manufacturers Assn., 96
National Geodetic Vertical Datum of 1929, 7

National Primary Drinking Water Regulations, 119
Naturally occurring organic compounds, 22
Net positive suction head (NPSH), 80
Neutron logging, 31

Open-pipe discharge, 141
Organic chemicals, 34–35
Orifice plate, 140

Parameter identification models, 74–75
Passive element system, 127
Permits, 35
Piping, 96
Plunger-type pumps, 88
Pollution control, 22
Polyphosphates
   See Glassy phosphates
Porosity, 4–5
Positive-displacement type pumps, 87
Precipitation, 1
   rates, 2
Pressure acidizing, 138
Pressure filters, 117–18
Professional services, 35
Propeller pumps, 87
Pump alignment, 96
Pump input horsepower, 93–94
Pump intake, 101
Pumping
   high-rate, 136
   tests, 140–43
Pumping rates, 66, 136
Pumps, 76
   continuous vs intermittent operation, 92–93
   installation, 96–98
   operating problems, 99–105
   operational limits, 95–96
   performance measurements, 93–95
   records, 120–21
   selection, 93–95
   specifications, 139–40
   types, 81, 84–85, 87–91

Quality management models, 74
Quantity management models, 74

Radial wells, 45
   yield, 69
Recharge, 63, 122–23
Reciprocating pumps, 88–90
Record keeping
   objectives, 120–21
Recovery method, 63–65
Resource management models, 71, 73–74
Reverse-circulation rotary method, 43–44
Rock, 3, 25, 123
   hydraulic conductivity, 12–13
Rock wells, 103–104
Rotary-displacement pumps, 87–88
Rotary drilling method, 43
Rotary-gear pumps, 88
Rotary pumps, 87

Saltwater intrusion, 22, 24
Sampling, 33–35
Sand pumping, 104–105, 138
Sands, 123
Sanitary construction, 53
Sanitary protection, 53–54
Saturated zone, 2
Schlumberger and Wenner array, 26–27
Sea level, 7
Seismic reflection and refraction, 25, 27–29
Sentinel wells, 33
Septic tank drain fields, 130–31
Single-point resistance logging, 30
Slime-forming bacteria, 145
Slot size, 50–52
Softening, 116–17
Soil zone, 2–3
Sounder, 142
Specific capacity, 138–39
Specific-capacity method, 65
Specific retention, 6
Specific yield, 6
Spontaneous potential (SP) logging, 30–31
Static discharge head, 70
Static head, 76
Static suction head, 76
Storage coefficient, 18–19, 55–56
Stovepipe method, 44
Straight-line solution, 62, 64–65
Submersible pumps, 85, 87
Subsidence
  See Land subsidence
Subsurface water, 2
Surface geophysical methods, 25–29
Surface water, 2
Surging, 136
Sustainable yield, 24
Synthetic organic compounds, 22

Tape method, 56
Test wells, 126–27
Theis nonequilibrium formula, 59–60
Topography, 9, 11
Total head, 7–8, 15
Total pumping head, 93
Transducers, 142–43
Transmissivity, 15–17, 55–56, 125
Treatment, 115–19
Tremie pipe, 50
Tubular wells, 135
Type-curve solution, 60–62, 64
Type curves, 58

Unconfined aquifers, 3, 59
Underground storage tanks, 130
Underreamed gravel-wall method, 46
Unit cube of material, 18
Unit prism of aquifer, 18
Unsaturated flow, 14–15
Unsaturated zone, 2–3, 23

Velocity, 52
Vibratory explosives, 104
Volute-type pumps, 81

Wastewater facilities, 130
Water demand, 144
Water-level change, 19
Water-level measurements, 56, 142–43
  analysis, 57–59
  data collection, 56–57
  hypothetical test, 57–59
  monitoring, 128–29
Water quality and contamination, 22, 24, 109, 112–14, 130–31
  chemical and physical characteristics, 109–111
  gases, 111–12
  management, 114
  monitoring, 33–35
  natural chemicals, 106–109
Water resources, vi
Water rights, 24
Water table, 3, 100–101
Water-table aquifers, 3
Water-table wells, 3
Weir box, 141
Well casings, 47 49
Well failures, 144–45
Well fields, 128
  availability and permeability, 129
  design, 65–67
  evaluation, 125–29
  interference, 66–67
  location, 129–31
  temperature, 129–30
Well inspection, 144
Well logs, 120
Well screens, 50–53, 101–102, 132–33
  selection, 133–34
Wells
  components, 47–53
  design, 131–35
  development, 135–39
  flow characteristics, 143
  losses, 68–69
  operation and maintenance, 144–45
  performance, 138–39
  sanitary protection, 53–54
  testing, 139–44
  type and construction, 39–46
Wetting agents, 137

Yield capacity, 131